Gaseous Matter

Joseph A. Angelo, Jr.

Facts On File
An Infobase Learning Company

This book is dedicated to the men and women of the United States Air Force—especially to my fellow airmen who served this great country with distinction and honor between 1967 and 1987.

Facts On File, Inc.
An imprint of Infobase Learning
132 West 31st Street
New York NY 10001

Library of Congress Cataloging-in-Publication Data
Angelo, Joseph A.
    Gaseous matter / Joseph A. Angelo, Jr.
        p. cm.—(States of matter)
    Includes bibliographical references and index.
    ISBN 978-0-8160-7607-9
    1. Gases—Popular works. 2. Matter—Properties—Popular works. I. Title.
QC161.A54 2011
530.4'3—dc22   2010034289

Facts On File books are available at special discounts when purchased in bulk quantities for businesses, associations, institutions, or sales promotions. Please call our Special Sales Department in New York at (212) 967-8800 or (800) 322-8755.

You can find Facts On File on the World Wide Web at http://www.infobaselearning.com

Excerpts included herewith have been reprinted by permission of the copyright holders; the author has made every effort to contact copyright holders. The publishers will be glad to rectify, in future editions, any errors or omissions brought to their notice.

Text design by Annie O'Donnell
Composition by Hermitage Publishing Services
Illustrations by Sholto Ainslie
Photo research by the author
Cover printed by Yurchak Printing, Inc., Landisville, Pa.
Book printed and bound by Yurchak Printing, Inc., Landisville, Pa.
Date printed: June 2011
Printed in the United States of America

10 9 8 7 6 5 4 3 2 1

This book is printed on acid-free paper.

Contents

Preface

The unleashed power of the atom has changed everything save our modes of thinking.

—Albert Einstein

Humankind's global civilization relies upon a family of advanced technologies that allow people to perform clever manipulations of matter and energy in a variety of interesting ways. Contemporary matter manipulations hold out the promise of a golden era for humankind—an era in which most people are free from the threat of such natural perils as thirst, starvation, and disease. But matter manipulations, if performed unwisely or improperly on a large scale, can also have an apocalyptic impact. History is filled with stories of ancient societies that collapsed because local material resources were overexploited or unwisely used. In the extreme, any similar follies by people on a global scale during this century could imperil not only the human species but all life on Earth.

Despite the importance of intelligent stewardship of Earth's resources, many people lack sufficient appreciation for how matter influences their daily lives. The overarching goal of States of Matter is to explain the important role matter plays throughout the entire domain of nature— both here on Earth and everywhere in the universe. The comprehensive multivolume set is designed to raise and answer intriguing questions and to help readers understand matter in all its interesting states and forms— from common to exotic, from abundant to scarce, from here on Earth to the fringes of the observable universe.

The subject of matter is filled with intriguing mysteries and paradoxes. Take two highly flammable gases, hydrogen (H_2) and oxygen (O_2), carefully combine them, add a spark, and suddenly an exothermic reaction takes place yielding not only energy but also an interesting new substance called water (H_2O). Water is an excellent substance to quench a fire, but it is also an incredibly intriguing material that is necessary for all life here on Earth—and probably elsewhere in the universe.

Matter is all around us and involves everything tangible a person sees, feels, and touches. The flow of water throughout Earth's biosphere, the air people breathe, and the ground they stand on are examples of the most commonly encountered states of matter. This daily personal encounter with matter in its liquid, gaseous, and solid states has intrigued human beings from the dawn of history. One early line of inquiry concerning the science of matter (that is, matter science) resulted in the classic earth, air, water, and fire elemental philosophy of the ancient Greeks. This early theory of matter trickled down through history and essentially ruled Western thought until the Scientific Revolution.

It was not until the late 16th century and the start of the Scientific Revolution that the true nature of matter and its relationship with energy began to emerge. People started to quantify the properties of matter and to discover a series of interesting relationships through carefully performed and well-documented experiments. Speculation, philosophical conjecture, and alchemy gave way to the scientific method, with its organized investigation of the material world and natural phenomena.

Collectively, the story of this magnificent intellectual unfolding represents one of the great cultural legacies in human history—comparable to the control of fire and the invention of the alphabet. The intellectual curiosity and hard work of the early scientists throughout the Scientific Revolution set the human race on a trajectory of discovery, a trajectory that not only enabled today's global civilization but also opened up the entire universe to understanding and exploration.

In a curious historical paradox, most early peoples, including the ancient Greeks, knew a number of fundamental facts about matter (in its solid, liquid, and gaseous states), but these same peoples generally made surprisingly little scientific progress toward unraveling matter's inner mysteries. The art of metallurgy, for example, was developed some 4,000 to 5,000 years ago on an essentially trial-and-error basis, thrusting early civilizations around the Mediterranean Sea into first the Bronze Age and later the Iron Age. Better weapons (such as metal swords and shields) were the primary social catalyst for technical progress, yet the periodic table of chemical elements (of which metals represent the majority of entries) was not envisioned until the 19th century.

Starting in the late 16th century, inquisitive individuals, such as the Italian scientist Galileo Galilei, performed careful observations and measurements to support more organized inquiries into the workings of the natural world. As a consequence of these observations and experiments, the nature of matter became better understood and better quantified. Scientists introduced the concepts of density, pressure, and temperature in their efforts to more consistently describe matter on a large (or macroscopic) scale. As instruments improved, scientists were able to make even better measurements, and soon matter became more clearly understood on both a macroscopic and microscopic scale. Starting in the 20th century, scientists began to observe and measure the long-hidden inner nature of matter on the atomic and subatomic scales.

Actually, intellectual inquiry into the microscopic nature of matter has its roots in ancient Greece. Not all ancient Greek philosophers were content with the prevailing earth-air-water-fire model of matter. About 450 B.C.E., a Greek philosopher named Leucippus and his more well-known student Democritus introduced the notion that all matter is actually composed of tiny solid particles, which are *atomos* (ατομος), or indivisible. Unfortunately, this brilliant insight into the natural order of things lay essentially unnoticed for centuries. In the early 1800s, a British schoolteacher named John Dalton began tinkering with mixtures of gases and made the daring assumption that a chemical element consisted of identical indestructible atoms. His efforts revived atomism. Several years later, the Italian scientist Amedeo Avogadro announced a remarkable hypothesis, a bold postulation that paved the way for the atomic theory of chemistry. Although this hypothesis was not widely accepted until the second half of the 19th century, it helped set the stage for the spectacular revolution in matter science that started as the 19th century rolled into the 20th.

What lay ahead was not just the development of an atomistic kinetic theory of matter, but the experimental discovery of electrons, radioactivity, the nuclear atom, protons, neutrons, and quarks. Not to be outdone by the nuclear scientists, who explored nature on the minutest scale, astrophysicists began describing exotic states of matter on the grandest of cosmic scales. The notion of degenerate matter appeared as well as the hypothesis that supermassive black holes lurked at the centers of most large galaxies after devouring the masses of millions of stars. Today, cosmologists and astrophysicists describe matter as being clumped into enormous clusters and superclusters of galaxies. The quest for these scientists is to explain how the observable universe, consisting of understandable forms of matter and energy, is also immersed in and influenced by mysterious forms of matter and energy, called dark matter and dark energy, respectively.

The study of matter stretches from prehistoric obsidian tools to contemporary research efforts in nanotechnology. States of Matter provides 9th- to 12th-grade audiences with an exciting and unparalleled adventure into the physical realm and applications of matter. This journey in search of the meaning of substance ranges from everyday "touch, feel, and see" items (such as steel, talc, concrete, water, and air) to the tiny, invisible atoms, molecules, and subatomic particles that govern the behavior and physical characteristics of every element, compound, and mixture, not only here on Earth, but everywhere in the universe.

Today, scientists recognize several other states of matter in addition to the solid, liquid, and gas states known to exist since ancient times. These include very hot plasmas and extremely cold Bose-Einstein condensates. Scientists also study very exotic forms of matter, such as liquid helium (which behaves as a superfluid does), superconductors, and quark-gluon plasmas. Astronomers and astrophysicists refer to degenerate matter when they discuss white dwarf stars and neutron stars. Other unusual forms of matter under investigation include antimatter and dark matter. Perhaps most challenging of all for scientists in this century is to grasp the true nature of dark energy and understand how it influences all matter in the universe. Using the national science education standards for 9th- to 12th-grade readers as an overarching guide, the States of Matter set provides a clear, carefully selected, well-integrated, and enjoyable treatment of these interesting concepts and topics.

The overall study of matter contains a significant amount of important scientific information that should attract a wide range of 9th- to 12th-grade readers. The broad subject of matter embraces essentially all fields of modern science and engineering, from aerodynamics and astronomy, to medicine and biology, to transportation and power generation, to the operation of Earth's amazing biosphere, to cosmology and the explosive start and evolution of the universe. Paying close attention to national science education standards and content guidelines, the author has prepared each book as a well-integrated, progressive treatment of one major aspect of this exciting and complex subject. Owing to the comprehensive coverage, full-color illustrations, and numerous informative sidebars, teachers will find the States of Matter to be of enormous value in supporting their science and mathematics curricula.

Specifically, States of Matter is a multivolume set that presents the discovery and use of matter and all its intriguing properties within the context of science as inquiry. For example, the reader will learn how the ideal gas law (sometimes called the ideal gas equation of state) did not happen overnight. Rather, it evolved slowly and was based on the inquisitiveness and careful observations of many scientists whose work spanned a period of about 100 years. Similarly, the ancient Greeks were puzzled by the electrostatic behavior of certain matter. However, it took several millennia until the quantified nature of electric charge was recognized. While Nobel Prize–winning British physicist Sir J. J. (Joseph John) Thomson was inquiring about the fundamental nature of electric charge in 1898, he discovered the first subatomic particle, which he called the electron. His work helped transform the understanding of matter and shaped the modern world. States of Matter contains numerous other examples of science as inquiry, examples strategically sprinkled throughout each volume to show how scientists used puzzling questions to guide their inquiries, design experiments, use available technology and mathematics to collect data, and then formulate hypotheses and models to explain these data.

States of Matter is a set that treats all aspects of physical science, including the structure of atoms, the structure and properties of matter, the nature of chemical reactions, the behavior of matter in motion and when forces are applied, the mass-energy conservation principle, the role of thermodynamic properties such as internal energy and entropy (disorder principle), and how matter and energy interact on various scales and levels in the physical universe.

The set also introduces readers to some of the more important solids in today's global civilization (such as carbon, concrete, coal, gold, copper, salt, aluminum, and iron). Likewise, important liquids (such as water, oil, blood, and milk) are treated. In addition to air (the most commonly encountered gas here on Earth), the reader will discover the unusual properties and interesting applications of other important gases, such as hydrogen, oxygen, carbon dioxide, nitrogen, xenon, krypton, and helium.

Each volume within the States of Matter set includes an index, an appendix with the latest version of the periodic table, a chronology of notable events, a glossary of significant terms and concepts, a helpful list of Internet resources, and an array of historical and current print sources for further research. Based on the current principles and standards in teaching mathematics and science, the States of Matter set is essential for readers who require information on all major topics in the science and application of matter.

Acknowledgments

I wish to thank the public information and/or multimedia specialists at the U.S. Department of Energy (including those at DOE headquarters and at all the national laboratories), the U.S. Department of Defense (including the individual armed services: U.S. Air Force, U.S. Army, U.S. Marines, and U.S. Navy), the National Institute of Standards and Technology (NIST) within the U.S. Department of Commerce, the U.S. Department of Agriculture, the National Aeronautics and Space Administration (NASA) (including its centers and astronomical observatory facilities), the National Oceanic and Atmospheric Administration (NOAA) of the U.S. Department of Commerce, and the U.S. Geological Survey (USGS) within the U.S. Department of the Interior for the generous supply of technical information and illustrations used in the preparation of this book set. Also recognized are the efforts of Frank K. Darmstadt and other members of the Facts On File team, whose careful attention to detail helped transform an interesting concept into a polished, publishable product. The continued support of two other special people must be mentioned here. The first individual is my longtime personal physician, Dr. Charles S. Stewart III, M.D., whose medical skills allowed me to successfully work on this interesting project. The second individual is my wife, Joan, who for the past 45 years has provided the loving and supportive home environment so essential for the successful completion of any undertaking in life.

Introduction

The history of civilization is essentially the story the human understanding and manipulation of matter. This book presents many of the important discoveries that led to the scientific interpretation of matter in the gaseous state. Readers will learn how the ability of human beings to relate the microscopic (atomic level) behavior of gases to their readily observable macroscopic properties (such as density, pressure, and temperature) helped transform the world.

Supported by a generous quantity of full-color illustrations and interesting sidebars, *Gaseous Matter* describes the basic characteristics and properties of several important gases, including air, hydrogen, helium, oxygen, and nitrogen. The three most familiar states of matter encountered on Earth are solid, liquid, and gas. Scientists call both gases and liquids fluids. Gases are quite common in the universe. Interstellar space has giant molecular clouds that contain mostly hydrogen and helium gas. Stars are gravitationally bound balls of very hot gases called plasma (the fourth state of matter). The giant planets of the solar system, Jupiter, Saturn, Uranus, and Neptune, are mainly gases but also possess frigid liquids that occur at great depths below their cloud tops. The inner terrestrial planets, Mercury, Venus, Earth, and Mars, are small, solid, rocky worlds, all of which (except Mercury) have relatively thin, shallow atmospheres.

Earth is a very beautiful and special planet. Two very thin skins of fluid cover the planet's surface, a low-density upper layer of fluid (composed mainly of gases), which scientists call the atmosphere, and a denser, lower layer of fluid (composed mainly of liquid water), which scientists call the oceans. Although often treated as separate physical entities (such as in the disciplines of atmospheric physics and oceanography), these two fluid layers are not completely distinct. Rather, water, gases, particulate matter, and energy transfer continuously across the interface between the fluid layers. These continuous exchanges have profound impacts on the development of weather systems (short-term impact) and climate change (long-term impact). *Gaseous Matter* discusses these important subjects as well as air pollution, the role of atmospheric ozone, and the consequences of carbon dioxide accumulation in the atmosphere due to fossil fuel burning.

Humankind's home planet is just large enough, has a sufficiently dense atmosphere, and is located at a favorable distance from its parent star, the Sun, so that liquids—primarily water (H_2O)—can exist and flow on its surface. Astrobiologists regard the presence of a protective atmosphere and liquid water on a planetary surface as essential for the development of life.

In general, a solid occupies a specific, fixed volume and retains its shape. A liquid also occupies a specific volume but is free to flow and assume the shape of the portion of the container it occupies. A gas has neither a definite shape nor a specific volume. Rather, it will quickly fill the entire volume of a closed container. Unlike solids and liquids, gases can be compressed easily. When temperatures are sufficiently high, plasma, the fourth state of matter, appears. As temperatures become very low and approach absolute zero, scientists encounter a fifth state of matter called the Bose-Einstein condensate (BEC).

Gaseous Matter presents the nature and scope of the science of fluids, highlights the most important scientific principles upon which the field is based, and identifies the wide range of applications that fluid science plays in almost all professional scientific and engineering fields. The role of gases in important natural phenomena, such as Earth's nitrogen cycle, is included.

The term *fluid* comprises both liquids and gases. On Earth, air is the most commonly encountered gas, while water is the most commonly encountered liquid. Both are essential for life. This volume focuses on gases, although scientific principles common to all fluids (both liquids and gases) are included for clarity and continuity. Long before the Scientific Revolution, which started in the middle of the 16th century in western Europe, ancient engineers and scientists such as Heron of Alexandria (ca. 20–80 C.E.) examined the behavior of fluids and developed simple devices that harnessed, controlled, and applied natural fluid science phenomena, such as steam, compressed air, and wind. Centuries later during the Scientific Revolution, pioneering scientists such as Galileo Galilei (1564–1642), Sir Isaac Newton (1642–1727), Evangelista Torricelli (1608–47), Blaise Pascal (1623–62), and Daniel Bernoulli (1700–82) began describing and predicting fluid behavior. Their technical efforts yielded important fluid science relationships for both liquids and gases.

Other early scientists used interesting experiments and mathematical relationships to help unlock nature's secrets. The ideal (perfect) gas equation of state is an important principle in the treatment of gases. This principle describes the physical relationship between the pressure (P), absolute temperature (T), and volume (V) of a gas. At low pressures and moderate temperatures, many real gases approximate ideal gas behavior quite well. As described in this volume, the important equation evolved after more than a century of careful, independent experimental work by the Anglo-Irish scientist Robert Boyle (1627–91), the French physicist Jacques Charles (1746–1823), and the French chemist Joseph-Louis Gay-Lussac (1778–1850). Modern scientists and engineers continue the intellectual tradition of describing complicated fluid behavior but are now assisted by the sophisticated, computer-based methodology called *computational fluid dynamics* (CFD).

It is important to realize that the science of fluids forms the cornerstone of modern civilization. Such diverse activities as power generation, air and surface transportation, space exploration, and modern medicine all depend on humankind's ability to understand and predict how gases and liquids behave under various physical conditions and circumstances.

Fluid science is the major branch of science that deals with the behavior of fluids (both gases and liquids) at rest *(fluid statics)* and in motion *(fluid dynamics)*. This scientific field has many subdivisions and applications, including *aerostatics* (gases at rest), *aerodynamics* (the motion of gases, including and especially air), *hydrostatics* (liquids at rest, especially water), and *hydrodynamics* (liquids in motion, including naturally flowing or artificially pumped water). *Gaseous Matter* concentrates on the interesting phenomena and scientific principles associated with gases, including aerodynamics, atmospheric physics, and kinetic theory. Special attention is given to the important gases in a person's daily life—air, oxygen, nitrogen, carbon dioxide, and natural gas (methane). This book also discusses the noble gases (helium, neon, argon, krypton, xenon, and radon) and several important but dangerous gases, such as chlorine and fluorine.

As previously mentioned, when scientists use the term *fluid,* they are referring to both liquids and gases. A fluid is a substance that, when in static equilibrium, cannot sustain a shear stress. A liquid will take the shape of and stay in an uncovered container, such as a bowl or coffee cup. In contrast, a gas easily takes the shape of the container but then escapes unless the container is tightly sealed.

An ideal fluid is one that has zero viscosity—that is, offers no resistance to shape change. Actual fluids only approximate this behavior. Viscosity is an important idea linked to internal fluid friction. This phenomenon is quite apparent to scientists who study the behavior of gases moving at high velocity.

Gaseous Matter describes how the German physicist Ludwig Prandtl (1875–1953) revolutionized fluid science when he presented his boundary layer concept in 1904. He discovered that while the bulk of a flowing fluid could be adequately treated using classical fluid science techniques, there was a thin region near the surface of a solid object where viscous effects dominated. His work led to a much better understanding of skin friction, a contribution of great importance to the emerging aeronautics industry.

From the beginning of human history, the theme of flight has permeated the myths, legends, art, literature, and religions of almost every culture. *Gaseous Matter* describes how science and engineering, especially an excellent understanding of gas dynamics, transformed humankind's long-cherished dream of soaring through the air like the birds into a modern reality. Modern aerospace technology has not only physically connected Earth's far flung inhabitants, it has allowed people to visit alien worlds.

Although fluids are very important in daily life, most people do not have a good understanding of the scientific principles that govern their behavior. To help remedy this circumstance, *Gaseous Matter* builds upon the three key ideas that form the overall architecture of fluid science. These important concepts are Newton's laws of motion, the continuity principle (indestructibility of flowing matter), and the conservation of energy. Scientists use other important concepts in their investigations of fluids. These additional notions include the idea of pressure (based on Pascal's principle of hydrostatics), the concept of buoyancy (based on Archimedes' famous principle), the ideal gas law, and the Bernoulli principle.

Historically, the scientific study of fluids concentrated first on the behavior of gases and then, as thermodynamics matured in the 19th century, incorporated the behavior of liquids under a variety of important physical conditions. Ancient peoples manipulated the flow of air, developed sailing ships, and eventually began using primitive windmills to grind grain. They experienced steam, but the scientific understanding of what actually happened when water vaporized awaited the development of thermodynamics in the 19th century. As part of the first Industrial Revolution, steam engines began to power factories and to serve as the principal motive force for trains and ships.

Gaseous Matter provides numerous examples of how fluid science principles and concepts govern different fields, including aeronautics and astronautics, astronomy, biology, engineering, medical sciences, chemistry, Earth science, meteorology, geology, oceanography, and physics. This volume has been carefully designed to help any student or teacher who has an interest in the behavior and application of gases and wishes to discover what gases are, how scientists measure and characterize them, and how the fascinating properties and characteristics of gases influenced the course of human civilization. The back matter of the book contains an appendix with a contemporary periodic table, chronology, glossary, and array of historical and current sources for further research. These should prove especially helpful for readers who need additional information on specific terms, topics, and events.

The author has carefully prepared the volume so that a person familiar with SI units (the international language of science) will have no difficulty understanding and enjoying its contents. He also recognizes that there is a continuing need for some students and teachers in the United States to have units expressed in the American customary system of units. Whenever appropriate, both unit systems appear side by side. An editorial decision places the American customary units first followed by the equivalent SI units in parentheses. This format does not imply the author's preference for American customary units over SI units. Rather, the author strongly encourages all readers to take advantage of the particular formatting arrangement to learn more about the important role that SI units play within the international scientific community.

CHAPTER 1

Gaseous Matter— An Initial Perspective

The ability of human beings to relate the microscopic (atomic level) behavior of gaseous matter to readily observable macroscopic properties (such as density, pressure, and temperature) transformed science and engineering. The great engineers of antiquity used gases effectively without possessing a detailed knowledge of their microscopic composition. The real breakthroughs in accurately describing the behavior of gases occurred in the 17th century as part of the Scientific Revolution. This chapter uses modern technology examples and historic anecdotes to describe some of the most important early efforts that led to the establishment of fluid science.

EARLY CONCEPTS OF MATTER

The early Greek philosopher Thales of Miletus (ca. 624–545 B.C.E.) was the first European thinker to suggest a theory of matter. About 600 B.C.E., he postulated that all substances came from water and would eventually turn back into water. Thales may have been influenced in his thinking by the fact that water assumes all three commonly observed states of matter: solid (ice), liquid (water), and gas (steam or water vapor). The evaporation of water under the influence of fire or sunlight could have provided Thales the notion of recycling matter.

Anaximenes (ca. 585–525 B.C.E.) was a later member of the school of philosophy at Miletus. He followed in the tradition of Thales by proposing one primal substance as the source of all other matter. For Anaximenes, that fundamental substance was air. In surviving portions of his writings, he suggested "Just as our soul being air holds us together, so do breath and air encompass the whole world." He proposed that cold and heat, moisture and motion made air visible. He further stated that when air was rarified, it became diluted and turned into fire. In contrast, the winds, for

THE WIND AND MYTHOLOGY

Modern scientists recognize that winds are the movement of air due to atmospheric pressure differentials. For most of human history, however, the winds were an unexplained but active and often very violent part of the physical world within which people lived. When a person experienced a gust of wind, the event generally coincided with an invisible, highly transitory atmospheric phenomenon. Ancient peoples often attributed such an unpredictable physical experience to the presence of a benevolent (or demonic) spiritual entity. They also recognized that air was necessary for life, so many of them assigned a spiritual significance to the wind and equated the unexplained natural phenomenon to the presence of an invisible life force. To reinforce this metaphysical (that is, beyond physical) interpretation, ancient languages sometimes used the same word for "breath" and "spirit." Even today, critics remark that a skillful artist has "breathed life" into a particular painting or that a gifted writer has "breathed life" into the characters of a fictional story.

The ancient Greeks (and later the Romans) personified and deified the winds, calling them the *anemoi* (*venti* in Latin). Using the general direction from which the wind came, the Greeks treated the four major winds as winged gods who inhabited Earth and brought about the weather conditions characteristic of the different seasons. For example, they called the north wind, *Boreas* (*Aquilo* in Latin). In Greco-Roman myth, Boreas was very strong and had a violent temper; he brought the cold, harsh winds of winter. *Notus* (*Auster* in Latin) was the south wind; he delivered the dry hot winds of midsummer and the crop destroying storms of late summer. *Eurus* (*Vulturnus* or *Eurus* in Latin) was the east wind, the wind god responsible for delivering rain and warm air. Finally, *Zephyrus* (*Favonius* in Latin) was the west wind. For the ancient Greeks, Zephy-

Anaximenes, represented condensed air. If the condensation process continued, the result was water, with further condensation resulting in the primal substance (air) becoming earth and then stones.

The Greek pre-Socratic philosopher Empedocles (ca. 495–435 B.C.E.) lived in Acragas, a Greek colony in Sicily. In about 450 B.C.E., he wrote the poem *On Nature*. In this lengthy work, of which only fragments survive, he introduced his theory of the universe, in which all matter is made up of four classical elements—earth, air, fire, and water—that

rus was the most gentle of the wind gods; his soft breezes signaled the arrival of spring.

Also within Greco-Roman myth, King Aeolus ruled the islands of Aeolia and served on behalf of the chief god, Zeus (Jupiter), as the regent (or keeper) of the winds. Readers of Homer's epic poem *The Odyssey* encounter King Aeolus just after Odysseus (Ulysses) and his men have escaped from the island of the Cyclops and ended up on the Aeolian Islands. Odysseus repays the king's hospitality by telling stories about the Trojan War and the adventures he and his men were experiencing as they attempted to return to Ithaca (a Greek island in the Ionian Sea). As they depart, King Aeolus presents Odysseus, a fellow king, with a leather bag in which he has placed all the stormy, troublesome winds. Aeolus wanted to provide Odysseus and his men a safe return to Ithaca. For several days, Odysseus remains steadfastly at the helm of the lead ship, guiding the other ships in his fleet back to Ithaca, but then the legendary king of Ithaca becomes exhausted and falls asleep—just as the gentle west wind brings the fleet within sight of home. Thinking that the gift from King Aeolus contained gold and silver, Odysseus's men decided to open it while their king slept. As howling winds rush out of the bag and push the fleet far away from shore, a startled Odysseus awakens and catches a brief glimpse of Ithaca receding on the horizon. The remainder of *The Odyssey* describes how Odysseus wanders for many more years before he eventually returns to his kingdom and his faithful wife, Queen Penelope.

Modern scientists use the term *aeolian* (sometimes spelled "eolian") to describe wind-related natural processes that shape a planet's surface, such as erosion. The personification of the wind also continues in the 21st century. Meteorologists and emergency planners assign human names to tropical storms, hurricanes, and cyclones.

periodically combine and separate under the influence of two opposing forces (love and strife). According to Empedocles, fire and earth when combined produce dry conditions, earth blends with water to form cold, water combines with air to produce wet, and air and fire combine to form hot.

In about 430 B.C.E., the early Greek philosopher Democritus (ca. 460–ca. 370 B.C.E.) elaborated upon the atomic theory of matter, initially suggested by his teacher Leucippus (fifth century B.C.E.). Democritus emphasized that all things consist of changeless (eternal), indivisible, tiny pieces of matter called *atoms*. According to Democritus, different materials consisted of different atoms that interacted in specific ways to produce the particular properties of a specific material. Some types of solid matter consisted of atoms with hooks, so they could attach to each other. Other materials, such as water and air, consisted of large, round atoms that moved smoothly past each other. What is remarkable about the ancient Greek theory of atomism is that it tried to explain the great diversity of matter found in nature with just a few basic ideas tucked into a relatively simple theoretical framework. The atomistic theory of matter fell from favor when the more influential philosopher Aristotle rejected the concept.

Starting in about 340 B.C.E., the famous Greek philosopher Aristotle (384–322 B.C.E.) embraced and embellished the theory of matter originally proposed by Empedocles. Within Aristotelian cosmology, planet Earth is the center of the universe. Everything within Earth's sphere is composed of a combination of the four basic elements: earth, air, fire, and water. Aristotle suggested that objects made of these four basic elements are subject to change and move in straight lines. However, heavenly bodies are not subject to change and move in circles. So Aristotle stated that beyond Earth's sphere lies a fifth basic element, which he called the aether (αιθηρ)—a pure form of air that is not subject to change. Finally, Aristotle suggested that he could analyze all material things in terms of their matter and their form (essence). Aristotle's ideas about the nature of matter and the structure of the universe dominated thinking in Europe for centuries until it was finally displaced during the Scientific Revolution.

Other ancient Greeks did more than contemplate the nature of matter. As some of the ancient world's most famous engineers, they learned through trial and error how to manipulate and use matter in a variety of intriguing ways.

ARCHIMEDES OF SYRACUSE

The Greek mathematician, inventor, and engineer Archimedes of Syracuse (ca. 287–212 B.C.E.) was one of the greatest technical minds in antiquity, if not all history. As a gifted mathematician, he perfected a method of integration that allowed him to find the surface areas and volumes of many bodies. This brilliant work anticipated by almost two millennia the independent co-development of the calculus by Sir Isaac Newton and Gottfried Wilhelm Leibniz (1646–1716) in the middle of the 17th century. In mechanics, Archimedes discovered fundamental theorems and physical relationships that described the center of gravity of plane figures and solids. These relationships lie at the very heart of modern mechanics and engineering dynamics.

He designed and constructed a variety of potent war machines in defense of his birth city of Syracuse against sieges by the Roman army during the Second Punic War. These military devices, not his brilliant mathematical contributions, made Archimedes famous in his lifetime. Science historians consider the absentminded Greek genius a mathematician comparable in brilliance to Newton, Leonhard Euler (1707–83), and Carl Friedrich Gauss (1777–1855).

Archimedes was born (ca. 287 B.C.E.) in the Greek city state of Syracuse on the island of Sicily. His father was an astronomer named Phidias, about whom very little is known. As a young man, Archimedes studied at the great library in Alexandria. Since he was a distant relative of Hieron II, king of Syracuse, Archimedes elected to return to his birth city and pursue his interests in mathematics, science, and mechanics. Although he personally regarded mathematics as a much higher level of activity than his efforts involving the invention of various mechanical devices, it was these engineering efforts and not his mathematics that earned him great fame.

In approximately 250 B.C.E., he designed an endless screw, later named the *Archimedes' screw*. This important fluid-moving device could efficiently remove water from the hold of a large sailing ship as well as pump water from rivers or lakes to irrigate arid fields. The device consists of a helical screw enclosed in a cylindrical casing that is open at both ends. The lowest portion of the pumping device is placed into the source body of water. As the helical screw turns, it scoops up a quantity of water. Subsequent revolutions of the screw move that quantity of water to other (higher) threads. Each revolution also adds a new scoop of water to the lowest thread. As the screw turns, water travels from thread to thread, until it reaches the top of the cylinder. Once there, it leaves the screw and pours into the intended destination.

At first, human or animal power turned the Archimedes' screw; eventually people learned how to harness wind power to operate this simple yet efficient device. The helical screw usually fits snuggly inside its cylindrical case but does not necessarily make a watertight seal. Save for the lowest thread, any water that seeps between threads during pumping is simply captured by a lower thread and hoisted upward again. Originally designed to help Egyptian peasants draw water out of the Nile River to irrigate their fields, the Archimedes' screw quickly found use throughout the Mediterranean basin and the Middle East. People still use the device to move water, especially in certain underdeveloped regions of Africa and Asia.

One of the most famous stories about Archimedes involves a challenge extended to him by Hieron II. The king of Syracuse wanted to determine whether a goldsmith made a requested crown by using the proper amount of pure gold, as instructed. There was some suspicion that the goldsmith had cheated the king by using a less expensive combination of silver and gold. Hieron II asked Archimedes to give him the right answer but without damaging the new crown in any way.

Archimedes thought for days about this problem. Then, as often happens to creative people, inspiration struck when least expected. As Archimedes stepped into a full bath, he observed the water spill over the sides. Immediately, he knew how to solve the king's intriguing problem. Jumping up out of the bath, he ran naked to the palace, enthusiastically shouting "Eureka!" This Greek exclamation means "I have found it!" Today, when an engineer or scientist experiences a similar intellectual breakthrough, they are said to have a "eureka moment."

What Archimedes had discovered in a flash of genius was the principle of buoyancy. Today, physicists refer to this phenomenon as the *Archimedes principle*. The principle states that any fluid applies a buoyant force to an object when that object is partially or completely immersed in it. The magnitude of this buoyant force equals the weight of the fluid the object displaces. Since the shape of an object is of no consequence, Archimedes was able to use this phenomenon to test the king's crown without damaging it. In antiquity (as now), silver was much cheaper than gold, so it was often used as a paste up substitute for gold by unscrupulous goldsmiths and jewelry merchants. Since silver has a lower density than gold, a crown fraudulently pasted up with silver in its interior would be bulkier (that is, have more volume) than a crown of identical weight made of only pure gold. If the volume of water displaced by the pure gold was equal to the volume of water displaced by the submerged crown, both could be assumed to have the same density and consist of identical material, namely pure gold. If the volume of water displaced by the crown was different than the volume of water displaced by an identical weight of pure gold, then there was some other metal, possibly lead or silver, in the crown along with the gold.

Archimedes carefully tested the suspicious crown and observed that the king's new crown displaced more water than the same weight of pure gold, so he concluded that the king's crown contained both gold and some other, less dense metal—most likely silver. The unwise goldsmith who tried to cheat King Hieron II confessed and was executed.

Today, scientists consider the Archimedes principle one of the basic laws of hydrostatics and aerostatics. This principle states that any object, entirely or partially submerged in a fluid (liquid or gas) is buoyed up by a force equal to the weight of the fluid displaced by the object. The Archimedes principle is the basis of lighter-than-air (LTA) vehicles. This principle explains why a balloon filled with helium or hot air rises up into Earth's atmosphere.

The ancient Greek engineer developed a number of important fundamental machines, including the lever and the compound pulley. With

Hot air balloons, like all lighter-than-air (LTA) vehicles, use the principle of buoyancy to rise into Earth's atmosphere. This photograph shows some of the more than 700 hot air balloons that took off in a mass ascension during the 2006 Balloon Festival in Albuquerque, New Mexico. *(U.S. Air Force)*

respect to the lever principle, Archimedes is reputed to have said: "Give me a place to stand on and I can move the Earth." Challenged by King Hieron II to move something very large, Archimedes developed a system of compound pulleys and levers and (according to legend) single-handedly pulled a fully loaded ship containing crew and cargo up out of the water and onto the shore with a single rope. Archimedes conducted other studies of force and motion. He discovered that every rigid body has a center of gravity, a single point at which the force of gravity appears to act on the body.

Many of Archimedes' surviving documents portray his wide-ranging interest in engineering and machines. These surviving works include *Theory of Levers, On Floating Bodies, On the Method of Mechanical Theorems,* and *The Water Clock.* He also wrote many other engineering-themed works that today are known only from cross references and prefaces in surviving books. Some of his missing works include *On Odometers, Winches, Hydroscopes, Pneumatics, On Balances or Levers, Centers of Gravity, Elements of Mechanics, On Gravity and Buoyancy,* and *Burning by Mirror.*

In addition to his genius for engineering, Archimedes was an incredibly gifted mathematician who resolved many important mathematical problems. For example, he made the most precise estimates of the value of π (the ratio of a circle's circumference to diameter) of his day. He was also a prolific writer in the field of mathematics, and some of his most important surviving works include *On the Sphere and Cylinder, Measurement of the Circle, On Spirals, On Tangential Circles, On Triangles, On Quadrangles,* and *On Conoids and Spheroids.* Despite achieving great fame through his mechanical inventions, Archimedes preferred to delve into mathematical problems, often getting absorbed for days and becoming oblivious to the world around him. Unfortunately, his lost-in-thought behavior would eventually prove fatal.

During his lifetime, Rome and Carthage fought for control of the Mediterranean basin. This power struggle resulted in a number of bloody conflicts called the Punic Wars. Rome waged three was against Carthage: the First Punic War (264–241 B.C.E.), the Second Punic War (218–201 B.C.E.), and the Third Punic War (149–146 B.C.E.). Carthage was defeated and totally destroyed in the Third Punic War, leaving Rome in complete control of the world around the Mediterranean Sea.

Archimedes became famous as a result of the many machines he developed for the defense of Syracuse during the First and Second Punic Wars. Specifically, he designed a variety of intricate machines to repulse attackers. Historians often place his military machines into three basic categories. First, there were the Archimedes' claws-cranes that could lift enemy ships up out of the water and smash them against the rocks. Next, there were a variety of catapults that could hurl rocks and other missiles over varying distances at enemy troops and ships. Finally, there was a collection of mirrors arranged to focus sunlight in such a way so as to set enemy ships on fire. This last development is open to a great deal of historic and technical speculation concerning its efficacy.

Whether Archimedes successfully used mirrors to set Roman ships on fire during the prolonged siege of Syracuse in the Second Punic War is not known for certain, but his other machines are known to have inflicted a great number of casualties on the attacking Romans. After Syracuse fell and the city was sacked, Archimedes was killed in 212 B.C.E. by a Roman soldier. The soldier slew the elderly Greek engineer despite standing orders from the Roman general Marcellus that the brilliant man be taken alive and treated with dignity.

The Roman historian Plutarch reported several accounts concerning the death of Archimedes. Two of these accounts are mentioned here. In the first account, Archimedes is murdered by a Roman soldier out of retribution, since the soldier wanted payback for so many of his comrades who were killed by Archimedes' machines. The other account suggests that as the city fell, a Roman soldier suddenly came upon Archimedes, who sitting on the ground drawing circles and other geometric figures in the sand. When told to move, the absent-minded Archimedes ignored the order and asked for time to finish the geometry problem. The impatient Roman gave him a fatal thrust with a short sword instead.

CTESIBIUS OF ALEXANDRIA

In ancient times, Archimedes' spectacular devices and discoveries generally overshadowed the technical accomplishments of another famous Greek inventor and engineer, Ctesibius of Alexandria (ca. 285–222 B.C.E.). Historians regard Ctesibius as the second most important engineer of antiquity. He made many contributions to the emerging disciplines of pneumatics, mechanics, hydraulics, and machine design.

Ctesibius published *On Pneumatics* in approximately 255 B.C.E. In this work, he discussed the elasticity of the air and suggested many applications of compressed air in such devices as pumps, musical instruments, and even an air-powered cannon (catapult). Science historians generally regard these efforts as the start of the science of pneumatics.

HYPERVELOCITY GAS GUNS

Vastly improving upon the early air cannon concept of Ctesibius, modern scientists and engineers use a variety of gas guns to study hypervelocity (up to about 5 miles per second [8 km/s]) collisions between small projectiles (typically 1/16 inch to ¼ inch in diameter [1.5 mm to 6.4 mm in diameter] and various solid target materials. Scientists define *hypervelocity* as a velocity in excess of about 1.25 miles per second (2 km/s), although the threshold of hypervelocity varies slightly from discipline to discipline.

NASA's Ames Vertical Gun Range (AVGR) was originally designed to conduct studies of lunar impact processes in support of the Apollo Project (lunar landing missions by astronauts). In 1979, NASA personnel transformed AVGR into a national hypervelocity research facility. Ballistic technologies, using a light gas gun or a gunpowder gun, allow researchers to shoot projectiles with velocities ranging from 0.6 to 4.3 miles per second (1 to 7 km/s) into various solid surface targets of interest in space exploration and aerospace engineering. The projectiles that scientists launch include spheres, cylinders, irregular shapes, and clusters of many small particles. These projectiles can be metallic (such as aluminum, copper, or iron), mineral (such as quartz or basalt), or glass. Today, this facility provides testing support for research in hypervelocity aerodynamics, impact physics, flow field structure, and chemistry.

The characteristic "energy flash" that occurs when a small projectile traveling at speeds up to 4.7 miles per second (7.6 km/s) impacts a solid surface at the Hypervelocity Ballistic Range at NASA's Ames Research Center in California. Such gas gun facility tests help aerospace engineers simulate what might happen when a piece of orbital debris (space junk) hits a spacecraft in orbit around Earth. *(NASA)*

The big gas gun at the Department of Energy's Lawrence Livermore National Laboratory (LLNL) in California creates shock waves that have pressures millions of times greater than the atmospheric pressure (nominally 14.7 psia [101 kPa]) that occur on Earth's surface. Such very high pres-

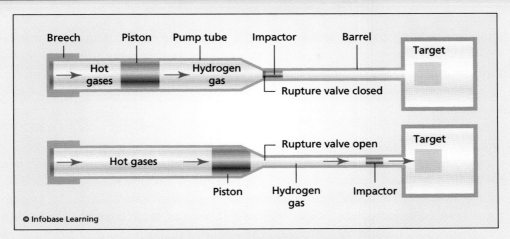

Breech   Piston   Pump tube   Impactor   Barrel

Target

Hot gases → Hydrogen gas

Rupture valve closed

Rupture valve open   Target

Hot gases →

Piston   Hydrogen gas   Impactor

© Infobase Learning

Operating principle for the two-stage, light gas gun capable of launching hyperve-locity impactors (projectiles) at research targets: (top) hot gases from exploding gunpowder drive piston compressing hydrogen gas in pump tube; (bottom) as pres-sure rises, highly compressed hydrogen gas bursts open rupture valve and acceler-ates impactor to the target. *(DOE/LLNL)*

sures (about 87×10^6 psia [600 GPa]) are typically encountered during intense chemical explosions, the detonation of nuclear weapons, as part of inertial confinement fusion experiments, or when a large meteorite hits Earth. LLNL is one of just a few institutions in the world with a major shock physics experi-mental program. Scientists have discovered that under shock conditions, they can induce novel configurations that give materials entirely new properties.

The centerpiece of LLNL's shock physics experimental program is a 65.6 foot-(20 m)-long, two-stage gas gun. (See diagram.) This gas gun consists of three major parts: a breech containing gunpowder, a pump tube filled with light gas such as hydrogen, and a barrel for guiding a high-velocity projectile (called an impactor) to the target. When the hypervelocity projectile hits the target, the impact produces a high-pressure shock wave.

Hot gases from the burning gunpowder drive a heavy piston down the pump tube, compressing the hydrogen gas. (See diagram.) Under compression, the pressure of the hydrogen gas increases until it reaches a value that breaks the rupture valve. Once the rupture valve breaks, the hydrogen gas hurls the impactor (projectile) down the gas gun's barrel and slams it into a target. The

(continues)

(continued)

chamber holding the target also contains a wide variety of diagnostic equipment that captures the data scientists need to measure equations of state, thermal and electrical conductivity, wave profiles, and the radiance (a combination of brightness and color) of the shocked sample. Scientists use a pyrometer to measure radiance and then translate radiance data into target temperature. Shocked-target temperatures in a gas gun experiment can exceed 13,120°F (7,273°C [7,000 K]).

The use of hydrogen gas produces the highest impactor velocities, which can range from approximately 2.48 miles per second (4 km/s) to 4.97 miles per second (8 km/s). A number of factors determine the terminal velocity of the impactor. These include the type and amount of gunpowder, the choice of light driving gas (hydrogen, helium, or nitrogen), the design break-pressure of the rupture valve, the diameter of the barrel, and the mass of the projectile. For example, LLNL scientists use projectiles with a mass of 0.033 lbm (15 g) and hydrogen as the light driving gas in the pump tube to achieve muzzle velocities up to 4.97 miles per second (8 km/s).

Modern engineers use pneumatic systems in a great many industrial and research applications. Typically, air (or an inert gas) is pumped by a centrally located compressor through pipes and flexible tubing to various pneumatic actuators or mechanical devices. In an automated system, an actuator is responsible for a specific action or sequence of actions. Today, engineers use actuators to move a factory robot's manipulator joints. Pneumatic actuators employ a pressurized gas, usually compressed air, to move the manipulator joint. When the gas is propelled by the pump through a tube to the particular joint, it triggers movement. Pneumatic actuators are inexpensive and simple, but their movement is not as precise as the smoother movements provided by electrical actuators.

Unfortunately, *On Pneumatics* along with all Ctesibius's other writings perished in the chaos of ancient times—primarily when great libraries such as the one in Alexandria were destroyed and their contents scattered. What is specifically known about the technical accomplishments of Ctesibius comes to us from other Greek inventors and engineers, such as

Heron of Alexandria (see next section) and the Roman military engineer and architect Marcus Vitruvius Pollio (ca. 80–15 B.C.E.). Ctesibius is credited with the invention of the siphon. He is also considered the creator of a small pipe organ called the *hydraulis,* which was supplied with air by a piston pump.

His greatest technical accomplishment was a vastly improved version of the water clock (clepsydra) of ancient Egypt. Ctesibius's improved water clock became the best timepiece in antiquity and remained unrivaled in accuracy until the 17th century. As a historic note, mechanical clocks were developed in medieval Europe. These devices were based on falling weights and proved to be more convenient than, but not as accurate as, Ctesibius's improved clepsydra. It was only the pendulum clock, introduced in the mid-17th century by the Dutch astronomer and physicist Christiaan Huygens (1629–95), that surpassed the accuracy of the water clock and ushered in a new era in timekeeping. Few mechanical devices have so dominated an area of technology for almost two millennia.

HERON OF ALEXANDRIA

Heron of Alexandria (also known as Hero) (ca. 20–80 C.E.) was the last of the great Greek engineers of antiquity. He invented many clever mechanical devices, including the device for which he is most commonly remembered, the aeolipile—a spinning, steam-powered spherical apparatus that demonstrated the action-reaction principle, which forms the basis of Sir Isaac Newton's third law of motion and the operating principle for modern reaction devices, such as gas turbines and jet engines.

Not much has survived about the personal life of Heron. Historians estimate that the Greek inventor was born in about 20 C.E. because his writings indicate that he observed a lunar eclipse, which was observable in Alexandria in 62 C.E. Heron had a strong interest in simple machines, mechanical mechanisms (such as gears), and fluid (hydraulic and pneumatic) systems. His inventions and publications reflect the influence of Ctesibius. Several of Heron's works have survived, including *Pneumatics* (written about 60 C.E.), *Automata, Mechanics, Dioptra,* and *Metrics*.

His most familiar invention is the aeolipile. He placed a hollow metal sphere on pivots over a charcoal grill-like device. When water placed inside the metal sphere was heated over the brazier, steam formed and

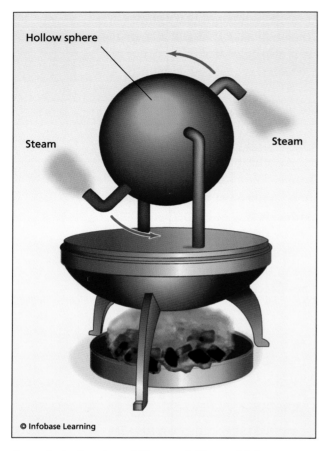

Hollow sphere

Steam

Steam

© Infobase Learning

An artist's drawing of Heron of Alexandria's aeolipile—a primitive, steam-powered reaction device *(Based on NASA artwork)*

escaped through the tubes, which acted like crude nozzles. The sphere would spin freely as steam escaped from two small opposing tubes connected to the sphere. This whirling sphere delighted children and became a popular toy. However, for some inexplicable reason, Heron never connected the action-reaction principle exhibited by the aeolipile with the basic concept of steam-powered devices for performing useful mechanical work.

The aeolipile is an example of a clever device invented well ahead of its time. Such devices sometimes need to be "rediscovered" or "reinvented" decades or even centuries later, when the social, economic, and/or technical conditions are right for full engineering development and application. Since the aeolipile embodies the action-reaction principle, it is the technical ancestor of modern power turbines, jet engines, and rocket engines. In the 20th century, steam turbines helped electrify the world, jet engines allowed modern military and commercial aircraft to swiftly reach distant points around the globe, and powerful rockets enabled the exploration of the solar system.

Despite this oversight regarding the aeolipile, Heron was a skilled engineer and creative inventor. He developed a variety of feedback control devices that used fire, water, and compressed air in different combinations. He designed a machine for threading wooden screws and constructed an automated puppet theater. He also receives credit for creating an early odometer, a primitive form of analog computer involving gears, spindles, weights, pegs, trays of sand, and ropes, and a compressed air (pneumatic) fountain.

TURBINES

A turbine is a machine that converts the energy of a fluid stream into mechanical energy of rotation. The working fluid used to drive a turbine can be gaseous or liquid. There are many types of turbines. A highly compressed gas drives an *expansion turbine,* hot combustion gases drive a *gas* (or *combustion*) *turbine,* steam (or other vapor) drives a *steam* (or *vapor*) *turbine,* water drives a *hydraulic turbine,* and wind spins a *wind turbine* (or *windmill*).

Except for wind- and water-powered turbines, a modern turbine typically consists of two sets of curved blades (or vanes) alongside each other. One set of vanes is fixed and the other set can move. Engineers space the moving vanes (called the *rotor*) around the circumference of a cylinder, which can rotate about a central shaft. They then attach the fixed set of vanes (called the *stator*) to the inside casing that encloses the rotor, or moving portion of the turbine.

In order to efficiently extract energy from the working fluid, gas and steam turbines usually contain a series of successive stages. Each stage consists of a set of fixed (stator) and moving (rotor) blades. The pressure of the working fluid decreases as it passes from stage to stage. Engineers increase the overall diameter of each successive stage to maintain a constant torque (or rotational effect) as the working fluid expands and loses pressure and energy.

(continues)

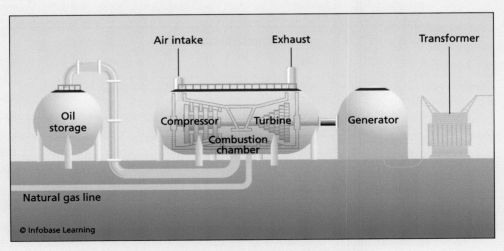

Air intake   Exhaust   Transformer

Oil storage   Compressor   Turbine   Generator

Combustion chamber

Natural gas line

© Infobase Learning

This is a diagram of a combustion turbine power plant. *(TVA)*

(continued)

Power system engineers design a combustion turbine (see diagram) to start quickly to meet the demand for electricity during peak periods. These turbines normally operate with natural gas as the fuel, although low–sulfur content fuel oil may also be used. The combustion turbine functions somewhat like a jet engine. However, no propulsive thrust is produced. Specifically, the system draws in ambient air at the front, compresses it, mixes the higher-pressure compressed air with fuel, and then combusts (ignites) the fuel-air mixture. As the hot combustion gases expand through the turbine, they spin (rotate) the turbine, which is located on a common shaft that is also connected to the compressor and to an electricity-producing generator. (In a jet engine, the combustion gases pass through a turbine and then expand through an exit nozzle to produce thrust.) After the hot gases pass through the power turbine and transform most of their energy content into the mechanical energy of rotary motion, the spent gases exhaust safely to the surroundings. The transformer at the power plant accommodates the long distance transmission of the electricity through the power grid.

HARNESSING THE WIND FOR WATER TRANSPORT

The development of wind-powered water transportation significantly influenced the trajectory of human civilization. Early peoples first discovered how to use simple rafts and dugout canoes to travel across inland waterways. Over time, the inhabitants of ancient civilizations learned how to construct ships that used both human muscles (oars) and wind power (sails) for propulsion. The Egyptians transported cargo along the Nile River using various types of barges and sailing ships. Then, as now, water transport offered an efficient way of moving large quantities of materials.

In about 1500 B.C.E., the Phoenicians emerged as the first great maritime trading civilization within the Mediterranean basin. From the coastal regions of what is now modern Lebanon, Phoenician sailors traveled across the Mediterranean in well-designed, human-powered sailing vessels, referred to by naval architects as *biremes*. The bireme had two sets of oars on each side of the ship and a large square sail. The ship's name results from a combination of *bi* (meaning "two") and *reme* (meaning "oars"). The Phoenicians were not only master shipbuilders and skilled traders, they also developed the early alphabet upon which present day alphabets are based. The ancient Greeks and later the Romans improved the design of the Phoenician bireme. The trireme (three rows of oars on each side) emerged as the dominant warship of the Mediterranean basin for centuries. Then, when the Dark Ages enveloped most of western Europe, Viking long ships began departing Scandinavian waters and prowling across the North Atlantic Ocean. Looking for trade or plunder, Norse sailors ventured far up many great European rivers and made daring sorties into the Mediterranean Sea.

As the nations of western Europe began their ambitious programs of worldwide exploration in the 15th century, their military and commercial interests encouraged the design of better sailing ships. For about the next 400 years, a parade of more efficient sailing ships carried commercial products, adventuresome people, and new ideas to all portions of the globe.

In 1794, the newly independent United States of America authorized the construction of six specially designed frigates for its fledgling navy. Reflecting characteristic Yankee ingenuity, these American sailing ships were larger and more heavily armed than the standard frigates of the day. The USS *Constitution* and her sister ships proved to be formidable opponents. Built in Boston of resilient live oak, the *Constitution*'s planks were up to seven inches (17.8 cm) thick. The American silversmith and patriot Paul Revere (1734–1818) forged the copper spikes and bolts that held the planks in place and the copper sheathing that protected the hull. The USS *Constitution* first put to sea in July 1798. In 1803, the sailing ship became the flagship for the American Navy's Mediterranean squadron. Ship and crew served with distinction against the Barbary States of North Africa, which were demanding tribute from the young nation in exchange for allowing American merchant vessels access to Mediterranean ports. During the War of 1812, the *Constitution* engaged and swiftly defeated the British frigate HMS *Guerriere.* The British sailors were so astonished that the shots from their cannons rebounded harmlessly off the *Constitution*'s hull that they gave the American ship the nickname "Old Ironsides." Today, the *Constitution* is the oldest commissioned warship still afloat and remains a powerful reminder of the historic connection between dominance of the sea and national security.

In 1807, a major breakthrough in water transport took place when the American engineer and inventor Robert Fulton (1765–1815) inaugurated commercial steamboat service. On August 14, 1807, the *Clermont,* Fulton's steamboat, made its journey up the Hudson River from New York City to Albany, demonstrating the great potential of steam-powered ships. A similar milestone in transportation occurred on January 17, 1955, when

On July 21, 1997, the USS *Constitution* (left) fires its guns in salute while underway under sail in Massachusetts Bay escorted by the frigate USS *Halyburton* (FFG 40) (center) and the destroyer USS *Ramage* (DDG 61) (right). The U.S. Navy's Blue Angels flight demonstration squadron passes overhead. Commissioned on October 21, 1797, the USS *Constitution* remains the world's oldest commissioned warship afloat. *(U.S. Navy)*

the world's first nuclear-powered ship, the submarine USS *Nautilus* (SSN-571), initially put to sea. As the *Nautilus* departed, its captain sent back the historic message "Underway on nuclear power."

Although well-designed sailing ships no longer form the backbone of military and merchant fleets, water transport remains a key component in the complex transportation system that supports the American economy. According to statistical data from the U.S. Department of Commerce, the United States is the world's leading maritime trading nation. More than 78 percent of American overseas trade by volume and 38 percent by value comes and goes by ship. American businesses rely on water transport to access suppliers and markets worldwide.

CHAPTER 2

Physical Characteristics of Gases

This chapter introduces the major physical characteristics of gases. Scientists use macroscopic properties (such as pressure and temperature) to describe and predict the thermodynamic behavior of physical systems. During the Scientific Revolution, European scientists such as Galileo Galilei, Sir Isaac Newton, Evangelista Torricelli, and Blaise Pascal investigated fluid behavior in terms of measurable physical quantities. Their early efforts began revealing fundamental fluid science relationships for both liquids and gases. Starting in the 18th century, other scientists such as Daniel Bernoulli built upon these pioneering efforts. The chapter describes how these scientists used interesting experiments and mathematical relationships to unlock nature's secrets. The modern understanding of gases on a macroscopic level emerged from this seminal era of science.

PHYSICAL CHANGE OF STATE

Scientists regard a physical change of state of a substance as one that involves a change in the physical appearance of that particular sample of matter, while the matter that experiences the physical change of state experiences no change in its atomic-level composition or chemical identity. Consider adding heat to an uncovered pot of water on a stove. The water eventually starts to boil and then experiences a physical change by

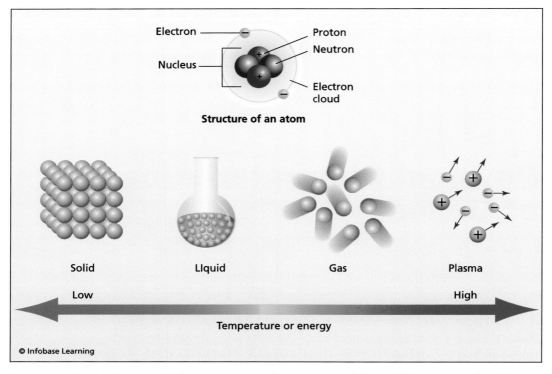

Electron
Proton
Neutron
Nucleus
Electron cloud

Structure of an atom

Solid   Liquid   Gas   Plasma

Low   High

Temperature or energy

© Infobase Learning

As energy is added to a solid, its temperature increases, and the solid becomes a liquid. Further addition of energy allows the liquid to become a gas. If the gas is heated a great deal more, its atoms break apart into charged particles (that is, electrons and positively charged nuclei), resulting in the fourth state of matter, plasma. *(Based on NASA-sponsored artwork)*

becoming a hot gas, commonly called *steam*. Scientists say that water experiences a physical change of state when it transforms to steam. Despite this change in physical state, steam retains the chemical identity of water.

Each of the three commonly encountered states of matter (solid, liquid, and gas) can change into either of the other two states by undergoing a change of state (or phase transition). The addition or removal of energy from the substance, usually as heat, facilitates such changes of state. The transitions are sensitive to the nature of the substance as well as its temperature and pressure. Scientists and engineers have developed a collection of technical terms to describe changes of state. They define *melting* as the change of a substance from the solid state to the liquid state; *freezing* is the opposite process. They define the *melting point* as the temperature for a specified pressure at which a substance transforms from the solid state to the liquid state. At this temperature, the solid and liquid states of a substance can coexist in equilibrium. The term *melting point* is synonymous with the term *freezing point*.

Scientists use the term *vaporization* to describe the process by which a substance in the liquid state becomes a gas. The term *evaporation* is also used. In the reverse process, called *condensation,* a substance in the gas (vapor) state becomes a liquid. When the evaporation rate of a liquid equals the condensation rate of the vapor, scientists say the liquid and vapor (gas) states of the particular substance are in equilibrium. They define the *boiling point* as the temperature at a specified pressure at which a substance in the liquid state experiences a change into the gas state. When a substance such as oxygen or nitrogen that is normally encountered as a gas on Earth experiences a change to the liquid state, scientists call the process *liquefaction.*

Sometimes a solid substance is at a temperature and pressure that allows it to transition directly into the gas (vapor) state. Scientists call this process *sublimation.* A frozen block of carbon dioxide (CO_2) at room temperature and one atmosphere pressure will transition (sublime) directly into gaseous carbon dioxide. Scientists refer to the reverse process as *deposition.* Under certain environmental conditions, water vapor in Earth's atmosphere transitions directly into the familiar solid known as snow. Once formed, snowflakes gently deposit themselves on the planet's surface.

Vapor pressure is an important thermodynamic property of a liquid substance. Engineers regard it as a useful measure of a liquid's inclination to evaporate. Volatile substances are those with high vapor pressures at normal temperatures and pressures. Chemists use the term *volatility* to describe the tendency of a liquid to vaporize. Consider a sealed flask of liquid with a sufficient unfilled volume (space) above the liquid's free surface. Under equilibrium conditions, a number of molecules in the liquid will escape across the free surface and occupy the open space as gas molecules. Scientists define the *vapor pressure* of a liquid as the partial pressure of the vapor over the liquid under equilibrium conditions at a specified temperature. The science of multistate (multiphase) substances is quite complicated. This book primarily focuses on substances in the gaseous (vapor) state.

Viewing matter at the microscopic (or atomic) scale, scientists understand now that gases (or vapors) form when there is a sufficient amount of energy in the system to enable its molecules (or atoms) to overcome most, if not all, of the attractive forces between them. A gas molecule moves about quite freely in any direction and, except for an occasional collision, has little interaction with its molecular neighbors. During the development of the kinetic theory of gases in the 19th century (discussed in chapter 4), scientists discovered that as the temperature of a gas increases, the amount of movement of its individual molecules (or atoms) also increases. Finally, since the molecules of a gas are typically quite far apart from each other, the gas can be easily squeezed (compressed) into smaller volumes. It is much harder to squeeze a solid or a contained liquid.

In general, a solid occupies a specific, fixed volume and retains its shape. A liquid also occupies a specific volume but is free to flow and assume the shape of the portion of the container it occupies. A gas has neither a definite shape nor a specific volume. Rather, it will quickly fill the entire volume of a closed container. Unlike solids and liquids that resist compression (or squeezing), gases are compressed easily. Compressibility is an important characteristic of all gases. Understanding how physical properties of a gas, such as pressure and temperature, change as it is being compressed (or expanded) allowed scientists to develop a very important tool called the ideal gas equation of state. (See chapter 3.)

The term *fluid* refers to both liquids and gases. On Earth, air is the most commonly experienced gas, and water is the most commonly encountered liquid. The great majority of people use the term *air* to describe the mixture of gases that make up Earth's atmosphere.

Near sea level, Earth's atmosphere contains the following approximate composition of gases (by volume): nitrogen, 78 percent; oxygen, 21 percent; argon, 0.9 percent; and carbon dioxide, 0.03 percent. There are also lesser amounts of many other gases, including water vapor and human generated chemical pollutants. During the Scientific Revolution, scientists investigated air as a gas, but they did not possess a detailed understanding of its basic chemical components. As discussed in subsequent chapters, pioneering combustion experiments and the aggressive search for new chemical elements enabled 18th- and 19th-century scientists to discover individual gases, such as nitrogen, oxygen, and carbon dioxide.

START OF MODERN FLUID SCIENCE

As part of the cultural and intellectual awakening that took place in Europe in the late Renaissance, scientists began describing and predicting fluid behavior. Their technical efforts yielded important fluid science relationships for both liquids and gases. Using measurable macroscopic properties such as pressure, volume, and temperature, early scientists discovered interesting physical relationships that helped explain and predict the behavior of fluids. Although an appreciation of the kinetic theory of gases was still centuries away, their dedicated research activities provided the foundation for the broad field of classical fluid science, sometimes called fluid mechanics. Modern scientists and engineers continue this intellectual tradition but are now assisted by the sophisticated, computer-based methodology called *computational fluid dynamics* (CFD).

It is important to realize that the science of fluids forms the cornerstone of modern civilization. Such diverse activities as weather forecasting, power generation, air transportation, maritime commerce, space exploration, modern medicine, and air pollution control all depend on humankind's ability to understand and predict how gases and liquids behave under various physical conditions and circumstances.

Fluid science is the major branch of science that deals with the behavior of fluids, both gases and liquids, at rest *(fluid statics)* and in motion *(fluid dynamics)*. This scientific field has many subdivisions and applications, including *aerodynamics* (the motion of gases, including and especially air), *hydrostatics* (liquids, including water) at rest, and *hydrodynamics* (liquids in motion, including naturally flowing or artificially pumped water). This chapter concentrates on the basic scientific properties and concepts associated with aerodynamics and gas dynamics.

A fluid is a substance that, when in static equilibrium, cannot sustain a shear stress. An ideal (perfect) fluid is one that has zero viscosity—that is, offers no resistance to shape change. Actual fluids only approximate this behavior. Viscosity is an important idea linked to internal fluid friction.

Three key ideas establish the overall architecture of fluid science. These important concepts are Newton's laws of motion, the continuity principle (indestructibility of flowing matter), and the conservation of energy. Scientists use other important concepts in their investigation of fluids. These additional concepts include the idea of pressure, specific volume, temperature, buoyancy, the ideal gas law, and the Bernoulli principle. Historically, the scientific study of fluids concentrated first on the behavior of gases and then, as thermodynamics matured in the 19th century, incorporated the behavior of liquids under a variety of important physical conditions.

SCIENTIFIC UNITS

Science is based on repeatable experiments, accurate measurements, and a system of logical, standardized units. The International System of units is the generally agreed upon coherent unit system now in use throughout the world for scientific, engineering, and commercial purposes. Universally abbreviated SI (from French Système International d'Unités), people also call it simply the metric system.

The contemporary SI system traces its origins back to France in the late 18th century. During the French Revolution, government officials decided to introduce a decimal measurement system. As a complement to the physical standards for length and mass carefully fabricated out of platinum, scientists used astronomical techniques to precisely define a second of time.

Scientists based the fundamental SI units for length, mass, and time—the meter (m) [British spelling, *metre*], the kilogram (kg), and the second (s)—on natural standards. Modern SI units still rely on natural standards and international agreement, but the new standards are ones that can be measured with much greater precision. Originally, scientists defined the meter as one ten-millionth the distance from the equator to the North Pole along the meridian of longitude nearest Paris. They denoted this standard of length with a carefully protected platinum rod. In 1983, scientists refined the SI definition of length with the following statement: "The meter is the length of the path traveled by light in vacuum during a time interval of 1/299,792,458 of a second." Scientists now use the speed of light in vacuum (namely 299,792,458 m/s) as the natural standard for defining the meter.

Other basic units in the SI system are the ampere (A) (as a measure of electric current), the candela (cd) (as a measure of luminous intensity), the kelvin (K) (as a measure of absolute thermodynamic temperature), and the mole (mol) (as a measure of the amount of substance). The radian (rad) is the basic SI unit of plane angle. The steradian (sr) is the basic SI unit of solid angle. There are also numerous supplementary and derived SI units, such as the newton (N) (as a measure of force) and the pascal (Pa) (as a measure of pressure).

Today, SI measurements play an important role in science and technology throughout the world. However, considerable technical and commercial activity in the United States still involves the use of another set of units, known as the United States (American) customary system of units.

NEWTON'S LAWS OF MOTION

Newton's laws of motion are the three fundamental postulates that form the basis of classical mechanics. He formulated these laws in about 1685 while studying the motion of the planets around the Sun. In 1687, Newton presented his work to the scientific community in *The Principia.*

Newton's first law of motion is concerned with the principle of inertia. It states that if a body in motion is not acted upon by an external force, the momentum remains constant. Scientists also refer to this law as the law of the conservation of momentum.

Newton's second law states that the rate of change of momentum of a body is proportional to the force acting upon the body and is in the direction of the applied force. A familiar statement of this law is the equation $F = m\,a,$ where F is the vector sum of the applied forces, m is the mass, and a represents the acceleration vector of the body.

Newton's third law is the principle of action and reaction. It states that for every force acting upon a body, there is a corresponding force of the same magnitude exerted by the body in the opposite direction.

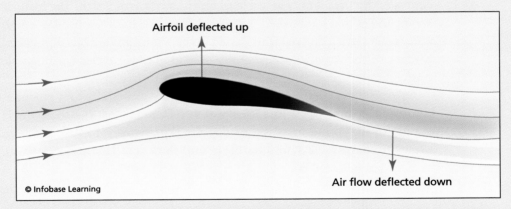

Airfoil deflected up

Air flow deflected down

© Infobase Learning

Newton's third law of motion (action-reaction principle) helps scientists explain how an airfoil generates lift. This diagram shows streamlines of air being deflected downward by the action of an airfoil; the reaction involves the aircraft's wing being pushed upward, creating lift. Scientists define a streamline as the line that describes the velocity direction of the fluid at each point along the flow. The velocity of the fluid may vary along a streamline. *(Based on NASA artwork)*

The American system traces its heritage back to colonial America and the weights and measures then being used throughout the British Empire in the 18th century.

One major advantage of SI units over American units is that SI units employ the number 10 as a base. Multiples or submultiples of SI units are easily reached by multiplying or dividing by 10. The inch (in) and the foot (ft) are the two commonly encountered units of length in the American customary system. Both originated in medieval England and were based upon human anatomy. The width of a human thumb gave rise to the inch; the human foot, the foot. Through an evolving process of standardization, 12 inches became equal to one foot, and three feet equal to one yard (yd).

The SI system uses the kilogram as the basic unit of mass, while the American customary system uses the pound-mass (lbm). The American system also contains another fundamental "pound" unit, the pound-force (lbf), which is actually a unit of force. Both "pounds" were related by an arbitrary decision made centuries ago to create a measurement system consisting of four basic (fundamental) units: length, mass, time, and force. To implement this decision, officials declared that "a one pound-mass object exerts (weighs) one pound-force at sea level." Technical novices and professionals alike all too often forget that this arbitrary pound equivalency is valid only at sea level on the surface of Earth. The historic arrangement that established the pound-force as a basic (or fundamental) unit rather than a derived unit (based on Newton's second law of motion) persists in the American customary system and can cause considerable confusion.

Scientists use SI units in their work to avoid ambiguity and potential confusion. Many people find it helpful to remember that the comparable SI unit for the pound-mass is the kilogram (kg), namely, 1 kg = 2.205 lbm, while the comparable SI unit for the pound-force is called the newton (N), namely, 1 N = 0.2248 lbf.

FORCE, ENERGY, AND WORK

The science of mechanics links the motion of a material object with the measurable quantities of mass, velocity, and acceleration. Through Newton's efforts, the term *force* entered the lexicon of physics. Scientists say a force influences a material object by causing a change in its state of motion. The concept of force emerges out of Newton's second law of motion. In its simplest and most familiar mathematical form, force *(F)* is the product of an object's mass (m) and its acceleration *(a)*—namely, $F = m\ a$. In his honor, scientists call the fundamental unit of force in the SI system the newton (N). A force of one newton accelerates a mass of one kilogram at the rate of one meter per second per second ($1\,N = 1\,kg\text{-}m/s^2$). In the American system, engineers define one pound-force (lbf) as the force equivalent to the weight of a one-pound-mass (1 lbm) object at sea level on Earth.

Physicists define *energy* (E) as the ability to do work. Within classical Newtonian physics, scientists describe *mechanical work* (W) as force *(F)* acting through a distance (d). The amount of work done is proportional to both the force involved and the distance over which the force is exerted in the direction of motion. A force perpendicular to the direction of motion performs no work. In the SI system, scientists measure energy with a unit called the joule (J). One joule represents a force of one newton (N) moving through a distance of one meter (m). This unit honors the British physicist James Prescott Joule (1818–89). In the American system, engineers often express energy in terms of the British thermal unit (Btu). (One Btu equals 1,055 joules.)

In classical physics, energy (E), work (W), and distance (d) are scalar quantities, while velocity *(v)*, acceleration *(a)*, and force *(F)* are vector quantities. A scalar is a physical quantity that has magnitude only, while a vector is a physical quantity that has both magnitude and direction at each point in space.

One of the most important contributions of Western civilization to the human race was the development of the scientific method and, through it, the start of all modern science. During the intellectually turbulent 17th century in western Europe, people of great genius began identifying important physical laws and demonstrating experimental techniques that helped humans everywhere better understand the physical universe.

CONCEPTS OF DENSITY

This section introduces the first of three familiar macroscopic physical properties: density, pressure, and temperature. By thinking about the atoms that make up different materials, scientists can now understand, quantify, and predict how interplay at the atomic (microscopic) level results in the physical properties that are measurable on a macroscopic scale.

To assist in more easily identifying and characterizing different materials, scientists devised the material property called density, one of the most useful macroscopic physical properties of matter. Solid matter is generally denser than liquid matter, and liquid matter denser than gases. Scientists define *density* as the amount of mass contained in a given volume. They frequently use the lower-case Greek letter rho (ρ) as the symbol for density in technical publications and equations.

Scientists use the density (ρ) of a material to determine how massive a given volume of that particular material is. Furthermore, equal volumes of different materials usually have different density values. Density is a function of both the atoms from which a material is composed as well as how closely packed the atoms are in the particular material. At room temperature (nominally 68°F [20°C]) and one atmosphere pressure, the density of some familiar materials is as follows: gold, 1,205 lbm/ft³ (19,300 kg/m³ [19.3 g/cm³]); iron, 493 lbm/ft³ (7,900 kg/m³ [7.9 g/cm³]); diamond (carbon), 219 lbm/ft³ (3,500 kg/m³ [3.5 g/cm³]); aluminum, 169 lbm/ft³ (2,700 kg/m³ [2.7 g/cm³]); and water, 62.3 lbm/ft³ (997.4 kg/m³)[0.997 g/cm³]. Like most gases at room temperature and one atmosphere pressure, oxygen has a density of just 0.083 lbm/ft³ (1.33 kg/m³ [1.33 × 10⁻³ g/cm³])—a value about 1,000 times lower than the density of most solid or liquid materials normally encountered on Earth's surface.

The maximum density of water (pure) occurs when the liquid's temperature is at 39.16°F (3.98°C). Upon freezing, water transforms to ice, which has a lower value of density—namely, 56.81 lbm/ft³ (910 kg/m³ [0.91 g/cm³]) at 32°F (0°C). This decrease in density upon freezing is the reason why ice floats on water, a very important yet unusual natural phenomenon. (Most materials have a higher density when solid than when liquid.)

Unlike rigid solids, fluids are materials that can flow, so engineers use pressure differentials to move fluids. They design pumps to move liquids (often treated as incompressible), while they design fans to move compressible fluids (gases). An incompressible fluid is assumed to have a constant value of density; a compressible fluid has a variable density. One of the interesting characteristics of gases is that, unlike solids or liquids, they can be compressed into smaller and smaller volumes.

Scientists use either the density or specific volume of a gas to describe its thermodynamic state. Engineers prefer to use the density of the gas as the intensive property associated with mass, because in many gas dynamic problems the mass of the gas varies from location to location.

Scientists know that the physical properties of matter are often interrelated—namely, when one physical property (such as temperature) changes, other physical properties (such as pressure or density) also change. As a direct result of the Scientific Revolution, people learned how to define the behavior of materials by developing special mathematical expressions, termed *equations of state*. Scientists created these mathematical relationships using both theory and empirical data from many carefully conducted laboratory experiments.

PRESSURE

Scientists describe pressure (P) as force per unit area. The most commonly encountered American unit of pressure is pounds-force per square inch (psi). In SI units, the fundamental unit of pressure is termed the pascal (Pa) in honor of Blaise Pascal (1623–62). The 17th-century French scientist conducted many pioneering experiments in fluid mechanics. One pascal represents a force of one newton (N) exerted over an area on one square meter—that is, $1\ Pa = 1\ N/m^2$. One psi is approximately equal to 6,895 Pa.

SPECIFIC VOLUME

In many fluid science problems, engineers find it necessary to include the density of the fluid. To assist in their computations, they define an intensive macroscopic property called the *specific volume* (usual symbol *v*). Specific volume is the volume per unit mass of a substance. It is the reciprocal of density and has the following units: ft^3/lbm in American units and m^3/kg in SI units.

An intensive thermodynamic property is independent of the mass involved in a particular problem; an extensive thermodynamic property varies directly with the mass and is dependent upon the actual size (or extent) of the system under study. The total volume (V) of a system is an example of an extensive property, while the specific volume *(v)* is an intensive property.

The use of intensive properties facilitates the development of thermodynamic property tables and the comparison of various physical processes with just a few well-understood thermodynamic laws and principles. For example, the density of dry air at 68°F (20°C) and one atmosphere pressure is 0.0755 lbm/ft³ (1.21 kg/m³). Therefore, the specific volume of dry air under the same conditions of temperature and pressure is 13.238 ft³/lbm (0.826 m³/kg). If engineers raise the pressure of this sample of air to 50 atmospheres while keeping the temperature at 68°F (20°C), the density becomes 3.777 lbm/ft³ (60.5 kg/m³) and the specific volume is 0.265 ft³/lbm (0.0165 m³/kg).

In the early part of the 17th century, European scientists remained puzzled about the notion of a vacuum and the possibility that Earth's atmosphere vanished at some point above the planet. They were strongly influenced in their thinking by Aristotle's teachings, which declared that

BLAISE PASCAL

The French physicist, mathematician, and philosopher Blaise Pascal (1623–62), performed important experiments involving fluids that led to the development of fluid science, especially the field of hydraulics. Pascal was born on June 19, 1623, in Clermont, France. Recognized early as a child prodigy, his father, a widower, moved the family to Paris in 1630 to further his young son's education.

Starting in about 1645, Pascal conducted a series of important experiments and developed several devices that applied the pressure of fluids. One of the main scientific products of his efforts was the important principle of hydrostatics, now called Pascal's principle. This principle states that any change in the pressure applied to a completely enclosed fluid is transmitted undiminished to all parts of the fluid and the enclosing container's walls. This basic principle governs the operation of hydraulic presses and elevators, air compressors, syringes, and similar fluid mechanics devices. He also confirmed and expanded upon the pioneering work of the Italian physicist Evangelista Torricelli concerning the decrease of atmospheric pressure with altitude and the existence of a vacuum. Pascal wrote strongly in defense of the scientific method and refuted the position of the French philosopher René Descartes (1596–1650) about the impossibility of a vacuum.

In 1654, a friend and gambler, the chevalier de Méré, asked Pascal to use his skills in mathematics to create an optimum strategy for a particular gaming house scenario. Pascal communicated with the French mathematician Pierre de Fermat (1601–65), and their correspondence and intellectual collaboration marks the start of probability theory. The real significance of this effort is that mathematics now began to address phenomena that were not precise and exact but rather statistical in nature. In addition to game theory, Pascal's work set the stage for such important areas of physics as statistical thermodynamics and quantum mechanics (based on the uncertainty principle of the German physicist Werner Heisenberg [1901–76]).

After a nearly fatal horse-drawn carriage accident at the Neuilly Bridge in 1654, Pascal turned his attention to philosophy and theology. Ill most of his life, he died in Port Royal on August 19, 1662.

nature abhors a vacuum *(horror vacui)*. The ancient Greek philosopher had speculated many centuries earlier that an empty space would simply suck any gas or liquid into itself in order to avoid being empty. The influential French mathematician and philosopher René Descartes (1596–1650) rejected the concept of action at a distance (as later incorporated by Sir Isaac Newton in his universal theory of gravitation) and thus could not accept the concept of a vacuum in space beyond Earth's atmosphere. So, Descartes devised an elaborate, though incorrect, vortex theory to explain celestial motions in the universe. Pascal challenged Descartes's opposition to the existence of a vacuum and conducted experiments in 1647 that expanded on the earlier atmospheric pressure measurements of the Italian scientist Evangelista Torricelli (1608–47).

In 1641, Torricelli received an invitation to visit Florence and serve as the blind, elderly Galileo's assistant. One of the important problems that challenged Galileo involved the difficulty that the pump makers of the grand duke of Tuscany encountered. They found that their suction pumps were not able to raise water beyond a height of about 32 feet (9.75 m). Before his death, Galileo suggested that Torricelli investigate the nature of this problem. When Galileo died in 1642, Torricelli became mathematician to the grand duke of Tuscany. Under the auspices of this position, he addressed the suction pump problem. Torricelli suspected the pressure of Earth's atmosphere might have something to do with the performance limitations experienced by the suction pumps.

In 1643, he devised a brilliant experiment. He took an approximately four-foot (1.2-m)-long glass tube that was sealed at one end and filled the tube with mercury. He then carefully placed the open end of the mercury-filled tube in a large basin of liquid mercury. With the tube secured in a vertical position, gravity caused the column of mercury in the tube to fall to a height of about 29.9 inches (76 cm) that left an empty space (sometimes called a Torricellian vacuum) at the top of the sealed glass tube. What Torricelli had invented was the mercury barometer, one of science's most important instruments. He also became the first scientist to create a sustained vacuum, thus smashing the long-held assumption that nature does not tolerate the existence of a vacuum.

At sea level, his instrument indicated that the pressure of Earth's atmosphere corresponds to a vertical column of mercury approximately 29.9 inches (760 mm) in height. Torricelli also observed that the height of the column of mercury in his barometer varied slightly from day to day and correctly concluded that this subtle change in height corresponded to a change in local atmospheric pressure. He further speculated that

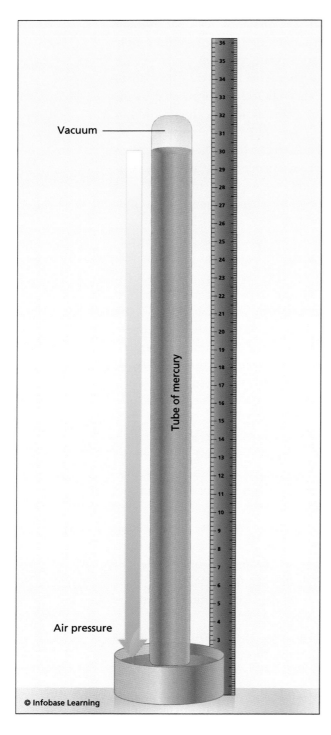

Vacuum

Tube of mercury

Air pressure

© Infobase Learning

such changes in atmospheric pressure might be the source of Earth's winds—the flow of air from regions of high pressure to regions of low pressure. In recognition of Torricelli's important discovery, scientists defined a special unit of pressure called the *torr*. One torr of pressure corresponds to a height of 0.254 inch (1 mm) of mercury in a barometer.

The pioneering work of Torricelli and Pascal guided other scientists in measuring and characterizing Earth's atmosphere. In 1654, the German scientist and politician Otto von Guericke (1602–86) provided a dramatic public demonstration of atmospheric pressure in the city of Magdeburg (and later elsewhere in Germany). No stranger to science, he had invented an air pump several years earlier. His famous experiment involved two hollow metal hemispheres, each about 1.64 feet (0.5 m) in diameter. By most historic accounts, once von Guericke had joined the hemispheres and pumped the air out, two teams of eight horses could not pull apart the assembled sphere, yet the hemispheres easily separated when von Guericke turned a valve and let air back into

This illustration shows a simple mercury barometer indicating that Earth's atmospheric pressure at sea level corresponds to a 29.92-inch-high (760-mm) column of mercury. *(Based on NOAA artwork)*

SUPERCRITICAL FLUID

Scientists define the *critical pressure* of a substance as the highest pressure under which the liquid and gaseous states of the substance can coexist. Similarly, the *critical temperature* is the temperature above which a substance cannot exist in the liquid state, regardless of the pressure. Above the critical temperature, the liquid smoothly transforms to the gaseous (vapor) state, and boiling disappears. The critical point occurs at the highest temperature and pressure at which liquid and gaseous states of a substance can coexist. Above the critical point, scientists call the fluid a *supercritical fluid*.

Within the supercritical fluid area shown in the accompanying diagram, the fluid exists in only one physical state and possesses both liquidlike and gaslike properties. Supercritical fluids have interesting thermodynamic prop-

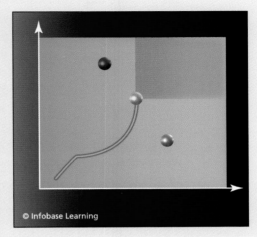

This plot describes the relationship between pressure and temperature for a typical substance in the liquid and gas state. Beyond the critical point, the substance becomes a supercritical fluid—exhibiting both gaslike and liquid-like properties. *(DOE/Pacific Northwest National Laboratory)*

erties. The physical characteristics of a supercritical fluid resemble those of dense gas, with solubilities approaching values experienced in the liquid state and dynamic viscosities intermediate between the values found in the liquid and gaseous states of the substance. The supercritical properties of carbon dioxide (CO_2), for example, make the abundant, nontoxic, and nonflammable substance attractive in industrial applications that require inexpensive solvents.

the evacuated sphere. The original metal hemispheres and air pump that von Guericke used in the Magdeburg experiment are now on display in the Deutsches Museum in Munich, Germany.

Scientists soon used the mercury barometer to measure the variation of atmospheric pressure with altitude or elevation above sea level. As they carried barometers up various mountains in Europe, they quickly discovered that Earth's atmospheric pressure decreased in a characteristic, measurable manner. An extrapolation of their results suggested that the pressure of Earth's atmosphere would eventually approach zero.

Atmospheric pressure plays an important role in many scientific and engineering disciplines. In an effort to standardize their research activities, scientists now use the following equivalent atmospheric pressure values for sea level: one atmosphere (1 atm) ≡ 760 mm of mercury (Hg) (exactly) = 29.92 in (Hg) = 14.695 psi = 1.01325×10^5 Pa.

One important feature of Earth's atmosphere is that the density in a column of air above a point on the planet's surface is not constant. Density and atmospheric pressure decrease with increasing altitude until both become negligible in outer space. (Chapter 5 discusses Earth's atmosphere in more detail.)

TEMPERATURE

While temperature is one of the most familiar physical variables, it is also one of the most difficult to quantify. Scientists suggest that on the macroscopic scale, temperature is the physical quantity that indicates how hot or cold an object is relative to an agreed upon standard value. Temperature defines the natural direction in which energy will flow as heat—namely, from a higher temperature (hot) region to a lower temperature (cold) region. Taking a microscopic perspective, temperature indicates the speed at which the atoms and molecules of a substance are moving.

Scientists recognize that every object has the physical property called temperature. They further understand that when two bodies are in thermal equilibrium, their temperatures are equal. A thermometer is an instrument that measures temperatures relative to some reference value. As part of the Scientific Revolution, creative individuals began using a variety of physical principles, natural references, and scales in their attempts to quantify the property of temperature.

In about 1592, Galileo attempted to measure temperature with a device he called the thermoscope. (Scientists customarily refer to Galileo Galilei by just his first name.) Although Galileo's work represented the first serious attempt to harness the notion of temperature as a useful scientific property, his thermoscope, while innovative, did not supply scientifically significant temperature data.

The German physicist Daniel Gabriel Fahrenheit (1686–1736) was the first person to develop a thermometer capable of making accurate, reproducible measurements of temperature. In 1709, he observed that alcohol expanded when heated, so he constructed the first closed-bulb glass thermometer with alcohol as the temperature-sensitive working fluid. Five years later, in 1714, he used mercury as the thermometer's working fluid. Fahrenheit selected an interesting three-point temperature reference scale for his original thermometers. His zero point (0°F) was the lowest temperature he could achieve with a chilling mixture of ice, water, and ammonium chloride (NH_4Cl). Fahrenheit then used a mixture of just water and ice as his second reference temperature (32°F). Finally, he chose his own body temperature (recorded as 96°F) as the scale's third reference temperature.

After his death, other scientists revised and refined the original Fahrenheit temperature scale, making sure there were 180 degrees between the freezing point of water (32°F) and the boiling point of water (212°F) at one atmosphere pressure. On this refined scale, the average temperature of the human body appeared as 98.6°F. Although the Fahrenheit temperature scale is still used in the United States, most of the other nations in the world have adopted another relative temperature scale, called the Celsius scale.

In 1742, the Swedish astronomer Anders Celsius (1701–44) introduced the relative temperature scale that now carries his name. He initially selected the upper (100-degree) reference temperature on his new scale as the freezing point of water and the lower (0-degree) reference temperature as the boiling of water at one atmosphere pressure. He then divided the scale into 100 units. After Celsius's death, the Swedish botanist and zoologist Carl Linnaeus (1707–78) introduced the present-day Celsius scale thermometer by reversing the reference temperatures. The modern Celsius temperature scale is a relative temperature scale in which the range between two reference points (the freezing point of water at 0°C and the boiling point of water at 100°C) are conveniently divided into 100 equal units, or degrees.

Scientists describe a *relative temperature scale* as one that measures how far above or below a certain temperature measurement is with respect to a specific reference point. The individual degrees, or units, in relative scale are determined by dividing the relative scale between two known reference temperature points (such as the freezing and boiling points of water at one atmosphere pressure) into a convenient number of temperature units (such as 100 or 180).

Despite considerable progress in thermometry in the 18th century, scientists still needed a more comprehensive temperature scale—namely,

RANKINE—THE OTHER ABSOLUTE TEMPERATURE

Most of the world's scientists and engineers use the Kelvin scale to express absolute thermodynamic temperatures, but there is another absolute temperature scale, called the Rankine scale (symbol R), that sometimes appears in engineering analyses performed in the United States—analyses based on the American customary units. In 1859, the Scottish engineer William John Macquorn Rankine (1820–72) introduced the absolute temperature scale that now carries his name. Absolute zero in the Rankine temperature scale (that is, 0 R) corresponds to −459.67 °F. The relationship between temperatures expressed in rankines (R) and those expressed in degrees Fahrenheit (°F) is:

$$T (R) = T (°F) + 459.67.$$

The relationship between the Kelvin scale and the Rankine scale is: (9/5) × absolute temperature (kelvins) = absolute temperature (rankines). For example, a temperature of 100 K is expressed as 180 R. The use of absolute temperatures is very important in science.

one that included the concept of absolute zero, the lowest possible temperature, at which molecular motion ceases. The Irish-born Scottish physicist William Thomson Kelvin (1824–1907), first baron of Largs, proposed an absolute temperature scale in 1848. The scientific community quickly embraced Kelvin's scale. The proper SI term for temperature is kelvins (without the word *degree*), and the proper symbol is K. Scientists generally use absolute temperatures in such disciplines as physics, astronomy, and chemistry; engineers use either relative or absolute temperatures in thermodynamics, heat transfer analyses, and mechanics, depending upon the nature of the problem. Absolute temperature values are always positive, but relative temperatures can have positive or negative values.

VISCOSITY OF AIR

Scientists use the property of viscosity as a measure of the internal friction, or flow resistance, of a fluid when the fluid is subjected to shear stress (τ). In the 17th century, Newton conducted relatively simple flow experiments that allowed him to make two important conclusions about fluid friction. His research still supports a basic understanding of the flow mechanics of real fluids.

Newton's first important observation was that the fluid does not move (or slide along) the fixed wall. This conclusion represents the important "no-slip" hypothesis in fluid mechanics. Since the fluid adheres to the surface of the stationary (fixed) wall, then the fluid velocity (v) is zero at this boundary. (For simplicity, vector notation is omitted here.) When scientists and engineers analyze frictional flow resistance in contemporary problems, they usually start by assuming that relative flow velocity at every point on the surface of a solid fixed boundary is zero. The German physicist Ludwig Prandtl (1875–1953) expanded Newton's work by introducing boundary layer theory in the early 20th century.

Newton's second conclusion involved his recognition that the resistive (viscous) force (F) on the moving upper wall is directly proportional to the relative velocity (v) and inversely proportional to the distance (y) separating the parallel walls. Assuming the shear stress (τ) equals the force (F) divided by the area (A), Newton and subsequent scientists were able to write: $\tau = F/A = \mu \, (v/H)$, where v is the velocity of the moving plate at the distance (y = H) from the fixed wall and μ is the dynamic viscosity coefficient. In more elaborate mathematical treatments of boundary layer phenomenon, differential calculus plays an important role. It allows scientists to express the shear stress for a fluid element at any point in the flow as follows: $\tau = \mu \, (\partial v/\partial y)$. The ratio $\partial v /\partial y$ represents the velocity gradient or rate of shearing strain and μ the constant of proportionality. Scientists refer to the term μ as the coefficient of viscosity or the dynamic viscosity. In American units, scientists quantify μ as (pounds-force-second)/ square foot, or ($lbf\text{-}s/ft^2$); in SI units, μ is expressed as (newtons-second)/ square meter, or ($N\text{-}s/m^2$). Since scientists define force per unit area as pressure, the dynamic viscosity also becomes pascal-seconds (Pa-s) in SI units. Physically, the dynamic viscosity represents the force that must be applied per unit area to permit adjacent layers of fluid to move with unit velocity relative to each other.

Scientists define kinetic viscosity (ν) as the dynamic viscosity divided by the fluid's density. In SI units, scientists express kinematic viscosity as m^2/s in SI units, and as ft^2/s in American customary units. At room temperature (nominally 68°F [20°C]) and one atmosphere pressure, water has a kinematic viscosity of approximately 1.076×10^{-5} ft^2/s (1.0×10^{-6} m^2/s), and air has a kinematic viscosity of 1.64×10^{-4} ft²/s (15.2×10^{-6} m²/s). In general, the viscosity of a liquid decreases with increasing fluid temperature; for a gas, the viscosity increases with increasing fluid temperature. For example, the kinematic viscosity of air at one atmosphere pressure and a temperature 1,500°F (816 °C [1,089 K]) is 15.1×10^{-4} ft²/s (140.3×10^{-6} m²/s). Aerodynamic heating due to fluid friction plays a major role in supersonic gas dynamics.

CHAPTER 3

The Rise of the Science of Gases

This chapter presents some of the fundamental concepts and governing physical principles that underlie the development of the modern science of gases. Starting with the important concept of a fluid behaving as a continuum, the scientific study of gases emerged from the late Renaissance period in western Europe and eventually expanded into the kinetic theory of gases in the late 19th century. The intellectual pathway was bumpy and contained many detours and blind alleys. Key discoveries along the way included the Bernoulli principle, the transition of alchemy into chemistry, the development of thermodynamics, the ideal gas concept, and, finally, the revival of atomism.

BASIC CONCEPTS

This section introduces several basic concepts used in the analysis of flowing fluids. If scientists can assume that the density of the fluid is constant, they can treat the fluid as incompressible. However, if the density of a fluid experiences significant changes, scientists must treat the fluid as compressible. Modern scientists understand that analyzing the high velocity flow of a compressible fluid (such as the hot combustion gases expanding through a jet engine's exhaust nozzle) is a much more complicated task than analyzing the flow of an incompressible fluid (such as water) through a pipe, but this was not always the case.

A fundamental idea in compressible fluid dynamics is that of the continuum. As part of the Scientific Revolution, scientists began treating a flowing fluid as a continuous material, or continuum. Today, scientists recognize that the approach is valid so long as the smallest volume of interest in a problem contains a sufficient number of atoms or molecules to make statistical averaging meaningful. The major advantage of the continuum approximation is that scientists can describe the complex behavior of innumerable individual atoms or molecules with a few macroscopic properties that describe (or quantify) the gross physical behavior of the substance. When scientists deal with compressible fluids and treat them as continua, they typically use the following macroscopic properties: density, pressure, temperature, viscosity, velocity, internal energy, enthalpy, entropy, and thermal conductivity.

As part of the continuum approximation, scientists and engineers assume that at any instant, every point of a fluid continuum has a corresponding fluid velocity vector. Instantaneous streamlines of flow are the curves that are tangent everywhere to the velocity vector. Streamlines are very useful in helping scientists visualize flow patterns. In steady flow, the streamlines are constant and represent the trajectories (path lines) taken by fluid particles. In unsteady flow, the streamlines change continuously.

The previous chapter discussed density, pressure, temperature, and viscosity. Brief mention will now be made of several other important macroscopic properties—properties that require an understanding of thermodynamics. The following thermodynamic properties are introduced here: internal energy, enthalpy, and entropy. More detailed discussions will occur wherever necessary to explain an important aspect of gas dynamics. Internal energy (symbol: u) is an intrinsic, macroscopic thermodynamic property interpretable through statistical mechanics as a measure of the microscopic energy modes (that is, molecular activity) of a system. This important property is associated with the first law of thermodynamics.

Enthalpy (symbol: h) is an intrinsic property of a thermodynamic system best described by its defining formula: $h \equiv u + P v$, where u is internal energy, P is pressure, and v is specific volume. Unfortunately, while very useful in the study of open systems and fluid flow problems, this thermodynamic property does not lend itself to a simple physical interpretation. However, as an intrinsic property of a thermodynamic system, specific enthalpy has the dimensions of energy per unit mass (Btu/lbm [J/kg]) and provides a macroscopic measure of the internal energy content and

FUNDAMENTALS OF THERMODYNAMICS

Thermodynamics is an elegant branch of physics based on several fundamental laws. Scientists and engineers use thermodynamics to describe how heat and mechanical energy (work) interact with substances in various systems. A *thermodynamic system* is simply a collection of matter and space (specified volume) with its boundaries defined in such a way that energy transfer (as work and heat) across the boundaries can be identified and understood easily. The *surroundings* represent everything else that is not included in the thermodynamic system. Scientists use the term *steady state* to refer to a condition in which the properties at any given point within a thermodynamic system remain constant over time. Neither mass nor energy accumulates (or depletes) in a steady state system.

A *closed system* is a system in which only energy but not matter can cross the boundaries. An *open system* can experience both matter and energy transfer across its boundaries. A *control volume* is a fixed region in space that is defined and studied as a thermodynamic system. Engineers often use a control volume in their analyses of open systems such as jet engines and gas turbine power plants.

The *zeroth law of thermodynamics* states that two systems, each in thermal equilibrium with a third system, are in thermal equilibrium with each other. This statement is actually an implicit part of the concept of temperature. The law is important in the field of thermometry and in the establishment of empirical temperature scales.

The first law is the conservation of energy principle. For a control mass (that is, a thermodynamic system of specified matter), engineers express the first law in the form of the following basic energy balance: energy output − energy input = change in energy storage.

The second law is an inequality asserting that it is impossible to transfer heat from a colder to a warmer system without the occurrence of other simultaneous changes in the two systems or the surroundings (environment). Another way of stating the second law is that the total change of entropy (ΔS) for an isolated system is greater than or equal to zero. Scientists express this particular statement mathematically as $(\Delta S)_{isolated} \geq 0$. An *isolated system* can experience neither matter nor energy transfer across its boundaries.

The third law states that the entropy of any pure substance in thermodynamic equilibrium approaches zero as the absolute temperature approaches zero. The law is important in that it furnishes a basis for calculating the absolute entropies of substances (either elements or compounds). These data then can be used in analyzing chemical reactions.

flow work performed on/by a fluid as it travels through a thermodynamic system.

Many technical papers and books have been written about the thermodynamic property called entropy (symbol: s). Entropy is an intrinsic thermodynamic property that serves as a measure of the extent to which the energy of a system is unavailable. Based on the second law of thermodynamics, specific entropy has the dimensions of energy per unit mass per degree of absolute temperature; typical units are Btu/(lbm-R) and J/(kg-K). In the late 19th century, as part of the development of kinetic theory, the Austrian physicist Ludwig Boltzmann (1844–1906) and the American scientist Josiah Willard Gibbs (1839–1903) introduced a statistical definition of entropy as a disorder or uncertainty indicator.

Thermal conductivity (symbol: k) is an intrinsic physical property of a substance that describes its ability to conduct heat (that is, transport energy) as a consequence of molecular motions in gases and liquids and as a combination of lattice vibrations and electron transport in solid substances. In 1822, the French mathematician Jean-Baptiste-Joseph Fourier (1768–1830) examined steady state heat transfer through a solid rod and provided a mathematical description of conduction heat transfer. He equated energy flow as heat as being proportional to the temperature gradient in the direction of heat flow and the area perpendicular to the heat flow. The constant of proportionality in his famous partial differential equation became known as the thermal conductivity (k) of the substance. Using mathematics, scientists express the one-dimensional, steady state version of this law as $q'' = -k\, \partial T/\partial x$, where q'' is the heat flux and $\partial T/\partial x$ is the temperature gradient. Fourier's law of heat conduction provided the first significant macroscopic measure of the rather elusive microscopic phenomenon of heat transfer by atomic motions.

In thermodynamics and gas dynamics, scientists make considerable use of the closed and open systems. A closed system is a thermodynamic system in which no transfer of mass takes place across the system's boundaries. However, energy in the form of heat and work may cross the system's boundaries. A frequently used closed mass system model involves the piston cylinder. The basic thermodynamic system illustrated here is the collection of working fluid (gas) contained in the piston-cylinder arrangement. As the piston pushes in (that is, does work on the gas), the gas contained in the cylinder experiences a decrease in volume and an increase in pressure. (Scientists assume the piston is properly sealed so the gas cannot escape while being compressed.) If the piston is both rigid and well-insulated, heat will not be able to leave the compressed gas, and its temperature will rise. Scientists use the term *adiabatic* to describe a process in which there is no energy transfer as heat across the boundaries of a thermodynamic system. They often treat the gas in the cylinder as an ideal gas, which in this example is experiencing an adiabatic compression. If the cylinder is cooled during the compression process in such a way as to keep the compressed gas at its initial temperature, then scientists say the gas is undergoing an isothermal compression.

The gas in the piston-cylinder example provides an excellent opportunity to discuss the very important state principle of thermodynamics. The state principle assumes that the number of independent properties needed to define the thermodynamic state of a system is equal to the number of possible work modes plus one. Scientists define a *simple substance* as one that has only one possible work mode. The gas in the piston-cylinder assembly has one work mode, namely, volume change (ΔV). This means that scientists need only two independent properties to describe the state of the gas while it is being compressed or expanded. They often select two of the following properties: temperature, pressure, or specific volume.

The concept of a pure substance is also important. Scientists define a *pure substance* as one that is homogeneous and maintains the same chemical composition in all its common physical states (that is, as a solid, liquid, or gas). Water is an example of a pure substance, since it maintains the same chemical composition whether it is ice, liquid water, or

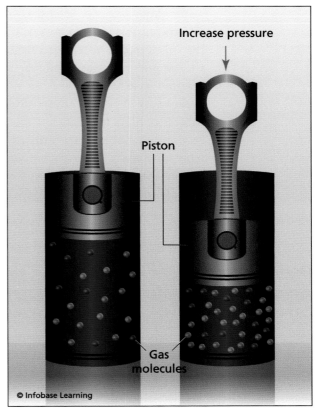

This illustration shows a fundamental closed mass system in thermodynamics consisting of the gas contained in a rigid-walled, piston-cylinder arrangement. *(EPA; adapted by author)*

water vapor (steam). Scientists often treat mixtures of gases such as air as a pure substance. However, if a gaseous mixture is cooled too much, the mixture can reach a sufficiently low temperature at which some of its component gases condense and become liquids. At that point, the original gaseous mixture can no longer be treated as a pure substance because its chemical composition has changed. The concept of a pure substance plays a significant role in thermodynamics. Unless noted otherwise, this book treats both air and water as simple substances and as pure substances.

The Swiss mathematician Leonard Euler (1707–83) and the Italian-French mathematician Joseph-Louis Lagrange (1736–1813) provided scientists the tools necessary to describe the motion of fluids in terms of a continuum. The method of Euler focuses on a fixed point in space and specifies the density, pressure, temperature, and so on, of the fluid particle that happens to occupy that point at each instant of time. Scientists prefer to use the Eulerian method when they solve problems involving fluid motion. The method of Lagrange involves the history of individual particles as they move through a system. In the Lagrangian method, the instantaneous density, pressure, temperature, state of stress, and so on, of a fluid particle of fixed identity are specified at each instant of time. Scientists use the Lagrangian description in solid mechanics.

When they apply the conservation of mass principle in classical fluid motion problems, scientists generally neglect relativistic effects and nuclear reactions. Invocation of this principle implies that the mass of the system is constant. Finally, scientists use Newton's second law of motion as the fundamental physical principle in fluid dynamics. The momentum theorem states that the net force acting instantaneously on the gas (or liquid) within a control volume equals the time rate of change of momentum within that control volume plus the excess of outgoing momentum flux minus the incoming momentum flux.

The assumption of a gaseous fluid behaving as a continuum fails whenever the mean free path of the molecules becomes comparable in size with the smallest significant physical dimension of the problem. For example, the continuum approach of classical gas dynamics is not appropriate for treating the highly rarified gases emitted by a rocket's nozzle as the rocket operates at very high altitude above Earth. For such problems, engineers must pursue a solution in terms of the microscopic description of matter found in kinetic theory.

BERNOULLI PRINCIPLE

The Swiss mathematician Daniel Bernoulli became an important pioneer in fluid science when he published the classic work *Hydrodynamica* in 1738. While exploring the relationship between pressure, density, and velocity in flowing fluids (especially water), he discovered that a moving fluid exchanges its kinetic energy for pressure. Scientists now call this important observation the Bernoulli principle.

To appreciate the true significance of Bernoulli's amazing insight, consider the steady state flow of an ideal (nonviscous but compressible) fluid through a section of pipe of varying geometries, as shown in the accompanying illustration. Since the flow is assumed to be steady, the mass flow rate ($\dot{m}$) at section 1 equals the mass flow rate at section 2. Scientists apply the continuity principle and define mass flow rate ($\dot{m}$) with the following equation: $\dot{m} = \rho\,A\,v$, where ρ is the fluid's density, A is the cross-sectional area of the pipe or conduit, and v is the fluid velocity. For a compressible (ideal) fluid, the continuity equation tells scientists that the product of the fluid's speed, density, and conduit's cross-sectional area is a constant. Scientist's express this very important relationship in fluid dynamics as $\rho_1\,A_1 v_1 = \rho_2\,A_2 v_2$.

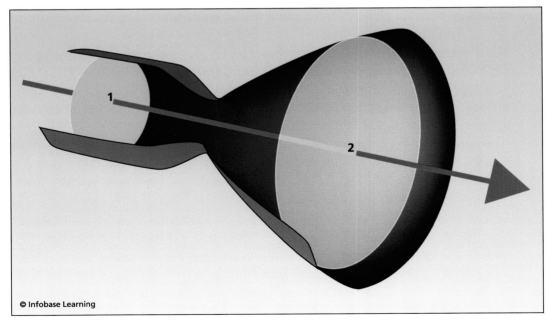

© Infobase Learning

Basic diagram that depicts flow through a variable area conduit *(NASA; adapted by author)*

Next, consider applying Bernoulli's brilliant observations to the frictionless, adiabatic, one-dimensional flow of a compressible fluid through the section of variable geometry pipe described in the accompanying figure. Scientists use the Bernoulli principle to write the following very important equation: $v \, dv + (1/\rho) \, dP = 0$, where v is velocity, ρ is density, and P is pressure. This is the Bernoulli principle mathematically expressed in differential form for a compressible fluid. It indicates that if the velocity of the fluid increases, its pressure decreases, and vice versa. However, scientists cannot integrate this equation unless they know how density varies with pressure.

This section has probably introduced enough assumptions and equations to make most readers somewhat uncomfortable. In truth, predicting the flow of real fluids that behave in nonideal conditions is very com-

COMPUTATIONAL FLUID DYNAMICS

Computational fluid dynamics (CFD) is the branch of science that integrates the use of numerical techniques, high-speed computers, property data, and complex theoretical and empirical equations to generate high-quality approximate solutions for some of the most challenging fluid science problems. At the heart of many CFD activities lie the Navier-Stokes equations. Claude-Louis-Marie-Henri Navier (1785–1836) was a French civil engineer who specialized in mechanics. The Irish mathematician Sir George Gabriel Stokes (1819–1903) investigated the flow of viscous fluids. The Navier-Stokes equations are based on Newtonian physics and describe the motion of fluids for a variety of interesting cases.

Scientists at the National Aeronautics and Space Administration (NASA) are using CFD technology to model the aerodynamics of advanced flight vehicles. The accompanying illustration is a computer-generated model of NASA's X-43a (Hyper-X) vehicle at Mach 7 flight test condition with engine operating. The CFD solution addresses internal (scramjet) and external flow fields, including the interaction between the engine exhaust and vehicle aerodynamics. CFD technology is one method of predicting a flight vehicle's performance. It also represents the best method aerospace engineers currently have for determining (prior to costly flight tests) a new vehicle's structural, pressure, and thermal design loads.

plicated. Today, most scientists and engineers depend on computational fluid dynamics (CFD) to get the job done right.

COMBUSTION HELPS SCIENTISTS UNDERSTAND GASES

This section describes how a study of combustion and heat helped scientists improve their understanding of gases. The phenomenon of heat had puzzled human beings since prehistoric times. Starting in the 17th century, alchemists and, later, pneumatic chemists used phlogiston theory to explain combustion and the process of oxide formation. The term *pneumatic chemist* applies to the scientists of the 17th through the early 19th centuries who studied chemical reactions and the physical behavior of

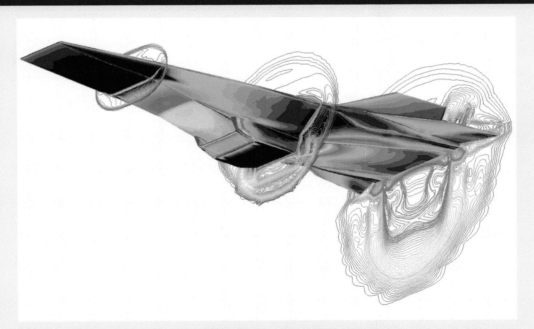

Computational fluid dynamics (CFD) image of NASA's X43 (Hyper-X) research vehicle at Mach 7 with engine operating. The computer-generated image illustrates surface heat transfer on the vehicle surface (red is the highest heating) and flow field contours at local Mach number. The last contour depicts the engine exhaust plume shape. *(NASA/DFRC)*

gases in an effort to understand the true nature and composition of matter.

Science historians identify the Flemish chemist and physician Johannes Baptista van Helmont (1579–1644) as the founder of pneumatic chemistry. Serving as a bridge between ancient alchemy and modern science, van Helmont stressed the importance of experiments in his chemical work, distinguished the existence of different gases in addition to atmospheric air, and coined the term *gas,* which he derived from the ancient Greek word for chaos.

In 1667, the German alchemist Johann Joachim Becher (1635–82) introduced a refined classification of substances, especially minerals. He replaced the air and fire portion of the four classical Greek elements with his own hypothesized three forms of earth. Becher called these new elemental substances terra lapidea, terra fluida, and terra pinguis. For him, terra lapidea represented the principle of inertness in matter, terra fluida the principle of fluidity (that is, the volatile or mercurial behavior of matter), and terra pinguis the principle of combustibility. His concepts were actively promoted by the 18th-century German chemist Georg Ernst Stahl (1660–1734), who renamed terra pinguis and called it *phlogiston*—from the ancient Greek word φλογιστος, meaning "fiery." Other 18th-century scientists also began using phlogiston theory to explain combustion and the process of oxide formation.

According to this now discarded theory, all flammable materials contained phlogiston, which was regarded as a weightless, colorless, and tasteless substance. During combustion, the flammable material released its phlogiston, and what remained of the original material was a dephlogisticated substance—a crumby or calcined residual material called *calx.* Phlogistonists assumed that when the phlogiston left a substance during combustion, it entered the air, which then became "phlogisticated air." Their pioneering combustion experiments with air in sealed containers revealed some very interesting yet initially inexplicable results.

Despite its shortcomings, during its intellectual reign, phlogiston theory played a significant role in the emergence of chemistry as a science and in the overall understanding of gases. For example, the Scottish chemist Daniel Rutherford (1749–1819) isolated one type of "phlogisticated air" that he called "noxious air" during combustion experiments in 1772. At the time, Rutherford was a student of the famous Scottish chemist and phlogistonist Joseph Black (1729–99), who had previously discovered carbon dioxide (CO_2) and named it "fixed air." Black was performing combustion experiments in which he burned a candle in a sealed glass container and watched the flame eventually extinguish itself. He knew one of the products that remained was his "fixed air" (or carbon dioxide), but he was not sure what the other remaining gas was. Hence, Black turned the problem over to Rutherford for further study.

Rutherford carefully extracted the carbon dioxide ("fixed air") using various chemical absorbers and then tested the remaining gas. Since it would not support life, Rutherford called it "noxious air." At the time, he and other phlogistonists thought that during combustion, phlogiston left the original air (which supported combustion) and deposited itself in the gaseous combustion products. Since carbon dioxide did not support combustion, the remaining "noxious air" was assumed to contain the postcombustion phlogiston—thus the basic name "phlogisticated air." Rutherford's left over so-called noxious gas is now known as *nitrogen,* the abundant atmospheric gas that does not support respiration and does not undergo combustion. Because of his experiments, Rutherford is generally credited with the discovery of nitrogen.

The French chemist Antoine Lavoisier (1743–94) performed carefully conducted mass-balance experiments in which he demonstrated that combustion involved the chemical combination of substances with oxygen. His work clearly dispelled the notion of phlogiston. In 1787, Lavoisier introduced the term *caloric* to support his own theory of heat, a theory in which heat was regarded as a weightless, colorless fluid that flowed from hot to cold. An incorrect concept called the conservation of heat principle served as one of the major premises in Lavoisier's caloric theory. Another important feature of this erroneous theory of heat was that the caloric fluid consisted of tiny, self-repelling particles that flowed easily into a substance, expanding the substance as its temperature increased.

In 1824, the French engineering physicist Sadi Carnot (1746–1832) published *Reflections on the Motive Power of Fire,* in which he correctly defined the maximum thermodynamic efficiency of a heat engine using caloric theory. Since steam engines dominated the Industrial Revolution, the thermodynamic importance of Carnot's work firmly entrenched the erroneous concept of heat as a fluid substance throughout the first half of 19th century. It took pioneering experiments by Benjamin Thompson (later known as Count Rumford) and James Prescott Joule to dislodge the caloric theory of heat.

In 1798, the British-American scientist Benjamin Thompson (1753–1814) published an important paper entitled *An Experimental Inquiry*

Concerning the Source of Heat Excited by Friction. Thompson had conducted a series of experiments linking the heat released during cannon boring operations in a Bavarian armory to mechanical work. He suggested that heat was the result of friction rather than the flow of the hypothetical fluid substance called caloric. The more Thompson bored a particular cannon barrel, the more heat appeared due to mechanical friction, an experimental result that clearly violated (and thus disproved) the conservation of heat principle of caloric theory.

In the early 1840s, the British physicist James Prescott Joule used his private laboratory to quantitatively investigate the nature of heat. He

LAVOISIER AND THE RISE OF MODERN CHEMISTRY

The French scientist Antoine-Laurent Lavoisier founded modern chemistry. He was born into a wealthy Parisian family in 1743 and educated at the Collège Marzin. Although trained as a lawyer, his interest in science soon dominated his lifelong pursuits. A multitalented individual, from the 1760s onward, he conducted scientific research while also being involved in a variety of political, civic, and financial activities.

In 1768, Lavoisier became a member of Ferme Générale, a private, profit-making organization that collected taxes for the royal government. Although this financially lucrative position opened up other political opportunities for the brilliant scientist, the association with tax collecting for the king would prove fatal. Lavoisier began an extensive series of combustion experiments in 1772, the primary purpose of which was to overthrow phlogiston theory. His careful analysis of combustion products allowed Lavoisier to demonstrate that carbon dioxide (CO_2) formed. He advocated the caloric theory of heat and suggested that combustion was a process in which the burning substance combined with a constituent of the air, a gas he called *oxygine* (a term meaning "acid maker").

As part of his political activities, Lavoisier received an appointment in 1775 to serve as one of the commissioners at the Arsenal in Paris. In this official capacity, he assumed responsibility for the production of gunpowder at the arsenal. Lavoisier's duties encouraged him to construct an excellent laboratory at the arsenal and to establish professional contacts with scientists throughout Europe.

One of Lavoisier's contributions to the establishment of modern chemistry was his creation of a system of chemical nomenclature that he summarized in discovered the very important relationship between heat and mechanical work. Although initially met with resistance, Joule's discovery of the mechanical equivalence of heat was eventually recognized as a meticulous experimental demonstration of the conservation of energy principle, one of the intellectual pillars of thermodynamics. His experimental results represent one of the great breakthroughs in science. In recognition of his accomplishments, the scientific community calls the SI unit of energy the joule (J).

Scientists and engineers now understand the physical mechanisms, both microscopic and macroscopic, that underlie the three basic modes

the 1787 publication *Method of Chemical Nomenclature*. Lavoisier's book promoted the principle by which chemists could assign every substance a specific name based on the elements of which it was composed. His nomenclature system allowed scientists throughout Europe to quickly communicate and compare their research, thus enabling a rapid advance in chemical science during that period. Modern chemists still use Lavoisier's nomenclature.

In 1789, Lavoisier published *Treatise of Elementary Chemistry*. Science historians consider it the first modern textbook in chemistry. In this book, Lavoisier discredited the phlogiston theory of heat and championed the caloric theory. He advocated the important law of conservation of mass and suggested that an element was a substance that could not be further broken down by chemical means. Lavoisier also provided a list of all known elements in this book, including light and caloric (heat), which he erroneously regarded as elemental, fluidlike substances.

When the French Revolution took place in 1789, Lavoisier initially supported it because he had previously campaigned for social reform. Throughout his life, he remained an honest though politically naïve individual. Lavoisier's privileged upbringing had socially insulated him from appreciating the growing restlessness of the peasants. Consequently, he did not properly heed the growing dangers around him and flee France. When the revolution became more extreme and transitioned into the Reign of Terror (1792–94), his membership in the much hated royal tax collecting organization (Ferme Générale) made him an obvious target for arrest and execution. All his brilliant contributions to the science of chemistry could not save his life. After a farcical trial, he lost his head to the guillotine in Paris on May 8, 1794.

of heat transfer: conduction, convection, and radiation. Conduction heat transfer takes place because of the motion of individual atoms or molecules. Convection involves the transfer of heat by bulk fluid motion. Radiation involves the transfer of energy by electromagnetic waves.

THE IDEAL GAS

The ideal (perfect) gas equation of state is an important principle in fluid science, especially in the treatment of gases. This principle states that the pressure (P), absolute temperature (T), and total volume (V) of a gas are related as follows: $P V = N R_U T$, where N is the number of moles of gas and R_U is the universal gas constant. For any gas, the universal gas constant (R_U) has the following value: 1,545.3 ft-lbf/(lbmol-R) in American units and 8,314.5 J/(kgmol-K) in SI units.

Engineers often prefer to express the ideal gas equation as $Pv = RT$, where v is the specific volume and R is the specific gas constant. At low pressures and moderate temperatures, many real gases approximate ideal gas behavior quite well. Assuming ideal gas behavior, air has a specific gas constant value (R) equal to 53.34 ft-lbf/(lbm-R), or 0.287 kJ/(kg-K), at one atmosphere pressure and 77°F(25°C).

The ideal gas equation evolved after a century of independent experimental work by the Irish-British scientist Robert Boyle, the French scientist Jacques-Alexandre-César Charles, and the French chemist Joseph-Louis Gay-Lussac. As the Scientific Revolution took hold in the 17th century, it produced an important transition in which the secret activities of alchemists gave way to the openly published and freely discussed material science activities of mechanical philosophers and pneumatic chemists, pioneering individuals who began exploring matter within the framework of the scientific method. Leading this new wave of intellectual inquiry was Robert Boyle, who published the book *The Skeptical Chymist* in 1661. In this work, he attacked the tradition of the four Greek classical elements (earth, air, water, and fire) and exposed the belief of alchemists in the philosopher's stone.

In the early 1660s, Boyle performed experiments concerning what he termed the "elastic" behavior of air. In 1662, he reported that there was a distinct relationship between the volume of air he had trapped in a U-tube shaped glass arrangement and its absolute pressure, which he carefully measured with a column of mercury. (Engineers define absolute pressure as the total pressure that a fluid exerts on the boundary [that is, walls] of a system.) Specifically, Boyle suggested the volume of gas (V) is inversely proportional to the pressure (P). In honor of his discovery, physicists refer to the equation P V = a constant for a gas at constant temperature as Boyle's law.

The French physicist Edmé Mariotte (ca. 1620–84) restated the same inverse relationship between pressure and volume for gases in his comprehensive essay *De la nature de l'air* (1679). To his credit, Mariotte recognized that the reciprocal relationship between pressure and volume only held for gases at constant temperature. He also used this relationship to estimate the height of the atmosphere. Because of this historic coincidence, scientific textbooks in France often refer to Mariotte's law of gases rather than Boyle's law.

On December 1, 1783, Jacques Charles constructed the first hydrogen-filled balloon and made an ascent to more than 10,000 feet (3,048 m). (Chapter 8 discusses lighter-than-air craft.) Between 1786 and 1787, he conducted experiments with several gases, including oxygen, nitrogen, carbon dioxide, and hydrogen. Charles observed, but did not publish, the fact that for a fixed quantity of each of these gases when held at constant pressure, the volume was inversely proportional to the temperature.

In 1802, another French scientist, Joseph-Louis Gay-Lussac, investigated the properties and behavior of gases and "rediscovered" Charles's law, namely that for a gas maintained at constant pressure, the volume varies inversely with its (absolute) temperature. Physicists refer to the reciprocal relationship V/T = a constant (for an ideal gas maintained at constant pressure) as Charles's law, Gay-Lussac's law, or sometimes the Charles–Gay-Lussac law. This relationship complements Boyle's law, and both were eventually combined into the ideal gas equation of state.

On August 24 1804, Gay-Lussac made the first of several of his daring ascents in a hydrogen-filled balloon. These flights ascended to altitudes above 23,000 feet (7,000 m) and enabled Gay-Lussac to measure the temperature, pressure, and humidity of the atmosphere as a function of height above sea level. His fellow aeronaut during the August ascent was the French physicist Jean-Baptiste Biot (1774–1862). While Gay-Lussac was busy investigating the properties of air during their balloon flight, Biot examined Earth's magnetism at high altitudes.

In 1805, Gay-Lussac observed that one volume of oxygen combined with two volumes of hydrogen to form water vapor. He pursued this line of investigation with other gases for several more years. Then, in 1808, Gay-Lussac reported his famous law of combining volumes. He concluded from his gas reaction experiments that the volumes of reactant gases at the same temperature and pressure are in ratios of small whole numbers. Gay-Lussac's work influenced both the British chemist John Dalton (1766–1844) and the Italian chemist Amedeo Avogadro (1776–1856). At the start of the 19th century, Dalton was also investigating combinations of gases and would revive atomism in order to explain his results. Three years after Gay-Lussac introduced his law of combining volumes, Avogadro boldly proposed in 1811 that equal volumes of any two gases at the same pressure and temperature contain the same number of molecules.

CHAPTER 4

Kinetic Theory of Gases

This chapter describes the important technical milestones that occurred in the last two centuries that resulted in the modern understanding of how gases behave. The centerpiece development took place in the second half of the 19th century and involved the kinetic theory of gases. Today, scientists use kinetic theory to successfully connect such useful macroscopic properties as pressure and temperature to the microscopic motion of individual atoms and molecules. However, the notion of a gas being composed of an enormous number of tiny atoms (either bound together as molecules or individually) was not completely embraced by the scientific community until the first decade of the 20th century. The chapter also introduces several other important concepts in contemporary gas dynamics, including diffusion, the boundary layer, supersonic speeds, and shock waves.

DALTON REVIVES INTEREST IN ATOMISM

In 1649, the French scientist and philosopher Pierre Gassendi (1592–1655) revisited the atomic theory of matter. His efforts served as an important bridge between the ancient Greek atomistic philosophy and the start of modern scientific atomism in the 19th century. Despite Gassendi's earlier efforts, science historians generally credit the British schoolteacher and chemist John Dalton (1766–1844) with the revival of atomism and the start of modern atomic theory. The first step in Dalton's revolutionary efforts occurred in 1801, when he observed that the total pressure of a mixture of gases is equal to the sum of the partial pressures of each individual

AVOGADRO'S BOLD HYPOTHESIS

In 1811, the Italian scientist Amedeo Avogadro reviewed Gay-Lussac's work and developed his own interpretation of the law of combining volumes. Avogadro hypothesized that equal volumes of any two gases at the same temperature and pressure contain the same number of atoms or molecules. Fellow scientists shunned Avogadro's bold hypothesis for many years—possibly because no one in the early 19th century had a clear understanding of the true nature of atoms and molecules.

Eventually, scientists did revisit his visionary hypothesis and then rewrote it in terms of the SI unit of substance, called the mole (mol). By international agreement (since 1971), scientists have defined a *mole* as the amount of substance that contains as many elementary units as there are atoms in 0.012 kg (12 g) of the isotope carbon-12. The precise number of atoms is a quantity called Avogadro's number (N_A). The elementary units in the definition of the mole may be specified as atoms, molecules, ions, or radicals. Scientists now treat Avogadro's number as a physical constant with the value 6.02×10^{23} mol^{-1}. According to Avogadro's bold hypothesis, 0.012 kg (12 g) of carbon-12 (that is, one mole) contains exactly 6.02×10^{23} atoms. While Avogadro never personally attempted to measure the value of the constant that now carries his name, his brilliant insight into the nature of matter played an important role in the development of the kinetic theory of gases.

By allowing them to measure the amount of gas present in a sample, Avogadro's number provided 19th-century scientists a useful starting point in developing the kinetic theory of gases. Chemists define the molar gas volume *(V_m)* as the volume of one mole of gas. To help standardize comparison between different gases, they usually compare volumes of gases at standard temperature (32°F [0°C {273 K}]) and standard pressure (one atmosphere). At standard pressure and temperature (STP), the molar gas volume is 22.41 liters per mole (L/mol) for an ideal gas. Real gases have values of molar volumes that agree within a few percent. At STP, helium (He) has a molar volume of 22.40 L; hydrogen (H_2), a value of 22.43 L; oxygen (O_2), a value of 22.39 L; carbon dioxide (CO_2), a value of 22.29 L; and ammonia (NH_3), a value of 22.09 L.

gas that makes up the mixture. Scientists now call this scientific principle Dalton's law of partial pressures.

In 1803, Dalton suggested that each chemical element was composed of a particular type of atom. He defined the atom as the smallest particle or unit of matter in which a particular element can exist. His interest in the behavior of gases allowed Dalton to quantify the atomic concept of matter. In particular, he showed how the relative masses or weights of different atoms could be determined. To establish his relative scale, he assigned the atom of hydrogen a mass of unity. Dalton's pioneering efforts revived atomic theory and introduced the concept of the atom into the mainstream of modern science.

Another important step in the emergence of atomic theory took place in 1811, when the Italian physicist Amedeo Avogadro (1776–1856) formulated his famous hypothesis. The hypothesis eventually became known as Avogadro's law. He proposed that equal volumes of gases at the same temperature and pressure contain equal numbers of *molecules*. At the time, neither Avogadro nor Dalton nor any other scientist had a clear and precise understanding of the difference between an atom and a molecule.

BASIC IDEAS ASSOCIATED WITH THE KINETIC THEORY OF GASES

This section discusses some of the basic ideas associated with the kinetic theory of gases. The physics of gases is one of the main subjects in thermodynamics, and the ideal gas equation of state is an important tool within thermodynamics. The kinetic theory of gases relates the macroscopic property of pressure to the collision of atoms (or molecules) with the walls of a container and the macroscopic property of absolute temperature to the kinetic energy of the atoms or molecules.

In 1738, the Swiss mathematician Daniel Bernoulli published a paper in which he provided a quantitative discussion of Boyle's law. This paper contained a rudimentary interpretation of kinetic theory. Bernoulli even suggested that particles (molecules) might move faster at higher temperature, but his paper drew very little attention from the scientific community. A little more than a century later, the Scottish physicist James Clerk Maxwell (1831–79) began publishing papers on the kinetic theory of gases. The Austrian physicist Ludwig Boltzmann (1844–1906) extended Maxwell's work by describing the velocity distribution for colliding gas molecules, and the Maxwell-Boltzmann distribution became an integral part of kinetic theory. (See next section.)

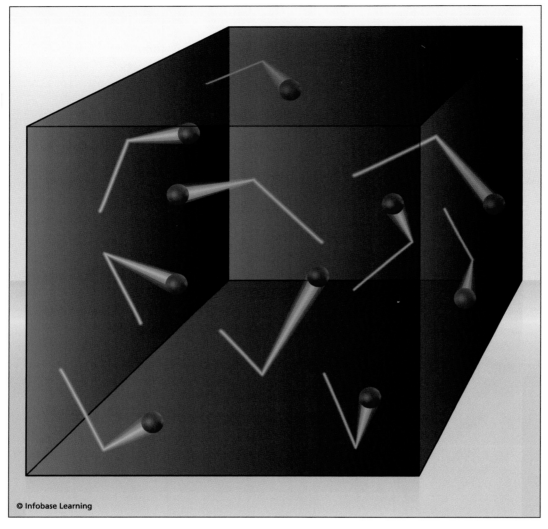

This illustration provides a microscopic interpretation of pressure. Within the kinetic theory of gases, the random motion of numerous gas molecules (or atoms) causes collisions with the container wall. During such elastic collisions, the molecules (or atoms) impart momentum to the walls, producing a force that can be measured. Scientists call this macroscopically observable force per unit area pressure. *(Based on EPA artwork)*

By linking microscopic atomic and molecular behavior to macroscopic properties of state such as pressure and temperature, the kinetic theory of gases proved very important in the full development of classical thermo-dynamics in the 19th century. Using kinetic theory, scientists determined that the average translational kinetic energy of a particle in an ideal gas is $3/2\ k\ T$, where k is Boltzmann's constant (discussed shortly) and T is the absolute temperature. Scientists also observed that the internal energy (U) of an ideal monatomic gas is given by the following expression: $3/2\ n\ R\ T$. Here, n is the number of moles of gas, and R is the universal gas constant. A monatomic ideal gas is composed of single atoms.

Scientists define internal energy (U) as a thermodynamic property that is interpretable through statistical mechanics as a measure of all the microscopic energy modes of a system. This important macroscopic property appears in a first law of thermodynamics energy balance for a closed system as follows: $Q - W = \Delta U$. Here, Q represents the heat added to the closed system, W represents the work done by (leaving) the system, and ΔU represents the overall change in the internal energy of the system. This deceptively simple equation relates the hidden behavior of an enormous number of atoms or molecules to readily observable macroscopic energy flows.

There are several key assumptions or postulates associated with the kinetic theory of an ideal gas. First, scientists assume that the gas consists of molecules (or atoms) that are of negligible size (volume) in comparison to the average distance between them. As an initial approximation, scientists such as Maxwell and Boltzmann ignored the volume occupied by the molecules (or atoms) in a gas.

Second, scientists postulate that the gas molecules (or atoms) move randomly in all directions and that they move in straight lines. As a result, any macroscopic properties of a gas, such as pressure, that depend on the motion of molecules will be isotropic in nature—that is, the same in all directions.

Third, scientists assume that colliding gas molecules (or atoms) experience elastic collisions. During an elastic collision, one particle can gain kinetic energy while another particle loses kinetic energy, but the total amount of kinetic energy is conserved.

Fourth, except for elastic collisions, scientists ignore (as negligible) any intermolecular (or interatomic) attractive or repulsive forces between the molecules (atoms) in a gas. Under this postulate of kinetic theory, gas molecules travel with undiminished speed in straight lines until they either collide with the container wall or else each other.

A fifth basic assumption of kinetic theory connects the macroscopic temperature of a gas to the individual motion of its molecules (or atoms). Scientists state that the average kinetic energy of a molecule (or atom) is proportional to the absolute temperature of a gas. The greater the kinetic energy of the molecules, the higher the absolute temperature will be. Scientists define the root mean square (rms) molecular speed ($\bar{v}_{RMS}$) as a type of average molecular speed equivalent to the speed of a molecule that possesses the average value of molecular kinetic energy. Based on the theoretical predictions of Maxwell, Boltzmann, and subsequent experimental demonstrations by other scientists, the $\bar{v}_{RMS}$ for hydrogen (H_2) molecules at 68°F (20°C [293 K]) and one atmosphere pressure is approximately 6,234 ft/s (1,900 m/s).

From kinetic theory, scientists know that if two ideal gases have the same temperature, their particles have the same average kinetic energy—expressed as ½ m $\bar{v}_{RMS}$. Even though two ideal gases may have the same temperature, the root mean square (rms) molecular speed ($\bar{v}_{RMS}$) can be different if the gas particles have different masses (m). Scientists can use kinetic theory to determine the speed of molecules in a container of air maintained at one atmosphere pressure and a temperature of 68°F (20°C [293 K]). As a first and very reasonable approximation, they assume the air in the container behaves like an ideal gas. Next, they assume the air is a mixture containing just nitrogen molecules (N_2) (which have a molecular mass of 28.0 atomic mass units) and oxygen molecules (O_2) (which have a molecular mass of 32.0 atomic mass units). Using kinetic theory, the scientists then determine that the root mean square (rms) molecular speed ($\bar{v}_{RMS}$) for nitrogen is 1,677 ft/s (511 m/s), while the $\bar{v}_{RMS}$ for oxygen is 1,568 ft/s (478 m/s). These results make perfectly good sense because even although both gases have the same temperature, their particles possess different masses, resulting in different molecular speeds for the same value of kinetic energy. (As described by the classical mechanics of Newton, the kinetic energy of a particle equals ½ m v^2, where m is the mass and v is the velocity.)

One very interesting conclusion is that even at "room temperature" conditions, the molecules in air (on average) move about randomly at very high speeds. For comparison, the speed of sound in air (a very important property in gas physics) under the same conditions of temperature and pressure is 1,122 ft/s (342 m/s).

LUDWIG BOLTZMANN

No discussion of kinetic theory is complete without mentioning the contributions of the brilliant, though troubled, Austrian physicist Ludwig Boltzmann (1844–1906). During his turbulent life, Boltzmann developed statistical mechanics as well as important thermoscience principles that enabled astrophysicists and astronomers to better interpret a star's spectrum and luminosity. In the 1870s and 1880s, he collaborated with the Austrian physicist Josef Stefan (1835–93), and together they developed the physical principle now called the *Stefan-Boltzmann law*. This law relates the total radiant energy output (luminosity) of a blackbody radiator (such as a star) to the fourth power of its absolute temperature.

Boltzmann was born on February 20, 1844, in Vienna, Austria. He studied at the University of Vienna, earning a doctoral degree in physics in 1866. His dissertation was supervised by Josef Stefan and involved the kinetic theory of gases. Following graduation, he joined his academic adviser as an assistant at the university.

From 1869 to 1906, Boltzmann held a series of professorships in mathematics or physics at a number of European universities. His physical restlessness mirrored the hidden torment of his mercurial personality, a personality that would swing him suddenly from a state of intellectual contentment into a mental state of deep depression.

His movement from institution to institution brought him into contact with many of Europe's most influential 19th-century scientists, including the German chemist Robert Wilhelm Bunsen (1811–99), the German theoretical physicist Rudolf Julius Emmanuel Clausius (1822–88), and the German physicist Gustav Robert Kirchhoff (1824–87). Among their many important contributions, Clausius developed the first comprehensive understanding of the second law of thermodynamics, while Bunsen collaborated with Kirchhoff to create the field of spectroscopy.

As a physicist, Boltzmann is best remembered for developing statistical mechanics and then pioneering its application in thermodynamics. His theoretical work helped scientists connect the properties and behavior of individual atoms and molecules (viewed statistically on the microscopic level) to the bulk properties and physical behavior (such as temperature and pressure) that a substance displayed when examined on the macroscopic level used in classical thermodynamics. One important development was Boltzmann's equipartition of energy principle. It states that the total energy of an atom or molecule is equally distributed, on an average basis, over its various translational kinetic energy modes. Boltzmann postulated that each energy mode had a corresponding degree of freedom. He further theorized that the average translational energy of a particle in an ideal gas was proportional to the absolute temperature of the gas. This principle provided a very important connection between the microscopic behavior of an incredibly large numbers of atoms or molecules and macroscopic physical behavior (such as temperature) that physicists could easily measure. The constant of proportionality in this relationship is now called *Boltzmann's constant* (symbol *k*). In SI units, the Boltzmann constant has a value of 1.380658×10^{-23} joules per kelvin (J/K).

Boltzmann developed his kinetic theory of gases independently of the work of the great Scottish physicist James Clerk Maxwell. Their complementary activities resulted in the *Maxwell-Boltzmann distribution,* a mathematical description of the most probable velocity of a gas molecule or atom as a function of the absolute temperature of the gas. The greater the absolute temperature of the gas, the greater the average velocity (or kinetic energy) of individual atoms or molecules. For example, consider a container filled with oxygen (O_2) gas at a temperature of 540 R (300 K). The Maxwell-Boltzmann distribution predicts that the most probable velocity of an oxygen molecule in that container is 1,296 ft/s (395 m/s). If the temperature of the oxygen increases to 2,160 R (1,200 K), then the Maxwell-Boltzmann distribution predicts that the most probable speed is near 2,625 ft/s (800 m/s).

In the late 1870s and early 1880s, Boltzmann collaborated with his mentor, Josef Stefan, and they developed a very important physical principle that describes the amount of thermal energy (heat) radiated per unit time by a blackbody. In physics, a blackbody is defined as a perfect emitter and perfect absorber of electromagnetic radiation. All objects emit thermal radiation by virtue of their temperature. The hotter the body, the more radiant energy it emits. By 1884, Boltzmann had finished his theoretical work in support of Stefan's observations of the thermal radiation emitted by blackbody radiators at various temperatures. The result of their collaboration was the famous Stefan-Boltzmann law of thermal radiation, a physical principle of great importance to physicists, engineers, and astronomers. The Stefan-Boltzmann law states that the luminosity of a blackbody is proportional to the fourth power of the luminous body's absolute temperature. The constant of proportionality for this relationship is called the Stefan-Boltzmann constant (symbol σ). The constant σ has a value of 0.1714×10^{-8} Btu/hr-ft^2-R^4 (5.669×10^{-8} W/m^2-K^4). The Stefan-Boltzmann law tells scientists that if the absolute temperature of a blackbody doubles, its luminosity will increase by a factor of 16.

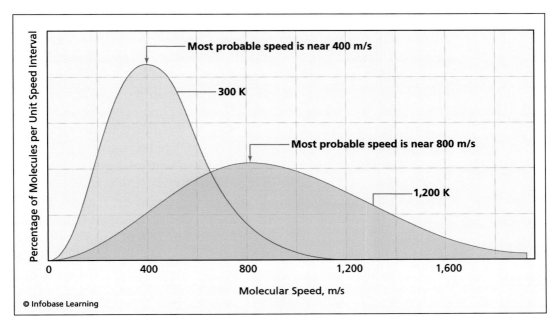

The Maxwell-Boltzmann curves portray the distribution of molecular speeds in oxygen, assumed to be an ideal gas at temperatures of 540 R (300 K) and 2,160 R (1,200 K). As the gas temperature increases, the distribution of molecular speeds broadens and the most probable speed increases. A speed of 400 m/s corresponds to 1,312 ft/s; 800 m/s corresponds to 2,625 ft/s. The rankine (R) is the unit of absolute temperature in the American system. *(Adapted by author from Cutnell and Johnson's Physics, Fifth Edition)*

The Sun and other stars closely approximate the thermal radiation behavior of blackbodies, so astronomers often use the Stefan-Boltzmann law to approximate the radiant energy output (or luminosity) of a stellar object. A visible star's apparent temperature is also related to its color. The coolest red stars (called stellar spectral class M stars) have a typical surface temperature of less than 6,300 R (3,500 K). However, very large hot blue stars (called spectral class B stars) have surface temperatures up to about 54,000 R (30,000 K) or more.

In the mid-1890s, Boltzmann related the classic thermodynamic concept of entropy (symbol S) that had recently been introduced by Clausius to a probabilistic measurement of disorder. In its simplest form, Boltzmann's famous entropy equation is $S = k \ln \Omega$. Here, he boldly defined entropy (S) as a natural logarithmic function of the probability of a particular energy state (Ω). The symbol k represents Boltzmann's constant. This important equation is even engraved as an epithet on his tombstone.

Toward the end of his life, Boltzmann encountered very strong academic and personal opposition to his atomistic views. Many eminent European scientists of the period could not grasp the true significance of the statistical nature of his reasoning. One of his most bitter professional and personal opponents was the Austrian physicist Ernest Mach (1838–1916), who was chair of the history and philosophy of science at the University of Vienna. Mach's bitter attacks forced Boltzmann to leave that institution in 1900 and move to the University of Leipzig.

Boltzmann persistently defended his statistical approach to thermodynamics and his belief in the atomic structure of matter. However, he could not handle having to continually defend his theories against mounting opposition in certain academic circles. While on holiday with his wife and daughter at the Bay of Duino near Trieste, Austria (now part of Italy), he hanged himself in his hotel room on October 5, 1906, as his wife and child were swimming. Was the tragic suicide of this brilliant physicist the result of a lack of professional acceptance of his work or simply the self-destructive climax of a lifelong battle with manic depression? No one can say for sure. Ironically, at the time of his death, other great physicists were performing key experiments that would soon prove his statistical approach to atomic structure not only correct but of great value to science.

DIFFUSION

Scientists define *diffusion* as the microscopic process by which atoms or molecules migrate from a region of higher mass concentration to one of lower mass concentration. During the process, the atoms or molecules travel on a zigzag journey, colliding with nearby particles until they slowly disperse within a given volume. Diffusion can occur in gases, liquids, or solids. In general, the rate of diffusion is fastest in gases, slower in liquids, and slowest in solids.

A thought experiment helps describe this interesting but complicated mass transfer phenomenon. However, any rigorous scientific treatment of diffusion lies well beyond the scope of this chapter and involves not only kinetic theory but the complex field of irreversible thermodynamics. In this simple thought experiment, imagine that a bottle of very fragrant perfume is opened and placed on a table located at one end of a large room. There is also a person sitting in a chair and reading at the opposite end of the room. Assume that the air is still (that is, no fans operating and no forced air circulation in the room) and that the room is at a uniform, standard value of temperature and pressure. Because of diffusion, the person sitting in the chair will eventually smell the perfume. On the microscopic level, the perfume molecules evaporate from the liquid, escape from the open bottle into the nearby air (creating a region of high perfume molecule concentration), and then slowly wander to other parts of the room due to numerous scattering collisions.

In diffusion studies, scientists refer to the host medium (such as the air in the previous thought experiment) as the *solvent,* while the diffusing substance is called the *solute.* Scientists are quick to point out that the general subject of mass transfer involves both mass diffusion on a molecular scale and bulk mass transport, as might take place during convection. Mass diffusion on a molecular scale can occur not only because of a concentration gradient, it may also occur due to a temperature gradient in a system. In the latter case, scientists call the process thermal diffusion. Sometimes a concentration gradient gives rise to a temperature gradient. To analyze such coupled phenomena mass transport problems, scientists resort to the field of irreversible thermodynamics.

The German physiologist Adolph Eugen Fick (1829–1901) developed a law of diffusion for fluids. A simple example will explain Fick's law. Consider a rectangular box that contains two gases called J and K that are initially separated by a removable partition. Gas J, on the left side of the partitioned box, could be carbon dioxide, and gas K, on the right, could be air. Once a scientist carefully removes the partition, the gases diffuse through each other until equilibrium is achieved. At equilibrium, the concentration of the gases will be uniform throughout the total volume of the rectangular box. Fick's law for this problem states that the mass flux ($\dot{m}$) of a constituent per unit cross-sectional area (A) is proportional to the concentration gradient *(C)* of that constituent. In a differential equation somewhat analogous to Fourier's law for conduction heat transfer, Fick's law yields the following mathematical expression for the one-dimensional mass flow rate of gas J ($\dot{m}_J$): $\dot{m}_J/A = -D\, \partial C_J / \partial x$. Here, $\dot{m}_J$ is the mass flux of gas J per unit time (lbm/s [kg/s]) while diffusing from a region of high initial concentration; D is the proportionality constant called the diffusion coefficient (ft^2/s [m^2/s]); and C_J is the mass concentration of component J per unit volume (lbm/ft^3 [kg/m^3]). The negative sign implies that molecules (or atoms) are diffusing away from regions

21 Minutes

U.S. government laboratories have developed rapid-response computer programs to help urban authorities react quickly and efficiently to the release of a chemical, biological, or radioactive contaminant by terrorists. The graphic shows a simulated contaminant plume spreading from New York City's Rockefeller Center. This rapid computational ability provides first responders the key plume dispersion information they would need to make effective decisions concerning population protection and decontamination activities. *(Naval Research Laboratory)*

of high mass concentration into regions of low mass concentration. This makes physical sense, because constituent molecules in a region of high concentration will, on average, experience more scattering collisions per unit time.

Diffusion is a complicated phenomenon. Today, scientists and security experts must be able to model the environmental transport (including diffusion, convection, and chemical reactions) of potentially lethal airborne contaminants intentionally released in a city by terrorists. Authorities must be able to rapidly predict the location of a contaminant plume and then quickly respond to the airborne chemical, biological, or radioactive threat.

URANIUM ENRICHMENT BY GASEOUS DIFFUSION

Uranium enrichment is the process of increasing the percentage of the uranium-235 isotope in a given amount of uranium, so that the uranium can then be used either as a fuel in a nuclear reactor (at a low to moderate enrichment) or as fissile material in a nuclear weapon (at very high enrichment). The enrichment process in the United States uses gaseous uranium hexafluoride (UF_6). In the past, large quantities of uranium were processed by gaseous diffusion. The only gaseous diffusion plant now in operation in the United States is located in Paducah, Kentucky.

Gaseous diffusion is based on the separation effect that arises from molecular effusion—the flow of gas through small holes. In a vessel containing a mixture of two gases, molecules of the gas with the lower molecular weight (here uranium-235, as opposed to uranium-238) travel faster and strike the walls of the vessel more frequently (relative to their concentration) than do the molecules that possess a higher molecular weight. Because the walls of the vessel are semipermeable, a larger number of the lower molecular weight molecules flow through the wall. Consequently, the gas that passes through the walls of the vessel is slightly enriched (a bit more plentiful) in the lower molecular weight isotope. Specifically, the isotope separation process is accomplished by diffusing uranium hexafluoride gas through a special porous membrane (barrier) and taking advantage of the different molecular velocities to achieve separation of uranium-235 from uranium-238.

Uranium hexafluoride (UF_6) is a chemical compound that consists of one atom of uranium combined with six atoms of fluorine. Engineers use UF_6 in uranium processing and enrichment because of its unique properties. Within a reasonable range of temperatures and pressures, UF_6 can be a solid, a liquid, or a gas. Of particular interest here is the fact that UF_6 becomes a gas at temperatures above 134°F (57°C) and one atmospheric pressure. Liquid UF_6 forms when the temperature exceeds 147°F (64°C) at a pressure greater than 1.5 atmospheres. UF_6 does not react with oxygen, nitrogen, carbon dioxide, or dry air. However, the compound does react with water or water vapor, forming corrosive hydrogen fluoride (HF) and a uranium-fluoride compound called uranyl fluoride (UO_2F_2).

COMPENSATING FOR REAL GAS BEHAVIOR

Because of its simplicity, the ideal gas equation of state is very useful, so scientists and engineers have developed special empirical relationships to widen its range of validity for real gases. One major assumption inherent in the ideal gas model is that gas molecules consist of infinitesimally small, hard, round spheres that do not exert any influence on each other (except for collisions) and that take up negligible volume. There is little intermolecular attraction for real gases at low pressures and high temperatures. Under such conditions, the gas molecules are actually far apart (on average), making the ideal gas approximation reasonable.

In the 1870s, the Dutch physicist Johannes Diderik van der Waals (1837–1923) pursued the problem and arrived at a new equation of state for gases known as the van der Waals equation that provided better agreement with experimental data, especially for gases under extreme conditions. The van der Waals equation improved late 19th-century kinetic theory by accounting for the weak attractions between gas molecules and for the finite volumes of the molecules. He received the 1910 Nobel Prize in physics for this important work.

The van der Waals equation and its significance in the physics of gases are briefly summarized here. In the 1870s, van der Waals proposed the following equation to approximate the behavior of gases: $(p + a/v^2)(v - b) = RT$. Here, p is pressure, v is specific volume, R is the specific gas constant (for a particular gas), T is absolute temperature, and a and b are constants. In developing this new equation of state for real gases, van der Waal replaced the pressure term (p) in the ideal gas equation with the term $(p + a/v^2)$ to account for weak but finite intermolecular forces. Then, to account for the fact that molecules in a gas of high density occupy a finite volume, van der Waals replaced the specific volume term v in the ideal gas equation with the term $(v - b)$, where b is a constant approximately related to the volume occupied by a unit mass in a dense (liquid) state.

Over the ensuing years since its introduction, scientists and engineers have empirically determined the constants a and b for many real gases, using best-fit statistical techniques for available experimental data.

MACH, SONIC VELOCITIES, AND SHOCK WAVES

This section provides a summary of sound waves, Mach numbers, and how shock waves are related to disturbances propagating through gases. Brief mention is also made of the phenomenon of sonic boom.

A very important property of any gas is the speed of sound through the gas. Physicists recognize that the speed of sound is actually the speed at which a small disturbance can travel (or propagate) through a particular medium. Human beings and other animals hear sound when a disturbance traveling through the air interacts with components of the ear, causing vibrations that are converted into nervous impulses that are then sent to the brain. The normal human ear can respond to sounds in the frequency range from approximately 20 to 20,000 hertz (Hz). Scientists define one hertz as one cycle per second. They call vibrations lower than this frequency range *infrasounds*, while they call vibrations above this frequency range *ultrasounds*.

Sound travels through gases, liquids, and solids at considerably different speeds. For air at standard conditions, the speed of sound is approximately 1,108 ft/s (338 m/s). By comparison, the speed of sound through water (at 68°F [20°C]) is about 4,862 ft/s (1,482 m/s), and through steel (at 68°F [20°C]) it is 19,492 ft/s (5,941 m/s). The speed of any mechanical wave (including a longitudinal sound wave) through a particular medium depends upon that medium's inertial properties and elastic properties. (A longitudinal wave involves oscillations that are parallel to the direction that the wave travels.) The inertial properties of a medium involve the storage of kinetic energy, while the elastic properties of the medium involve the storage of potential energy.

Disturbances are transmitted through a gas as a result of collisions between the randomly moving molecules or atoms of the gas. Scientists often treat the transmission of a small disturbance through a gas as an isentropic process. (In thermodynamics, an isentropic process is a constant entropy process—that is, a reversible, adiabatic process.) In an isentropic process, the conditions in the gas are the same before and after the disturbance passes through. Because the speed of transmission of the disturbance depends on molecular collisions, the speed of sound depends on the thermodynamic state of the gas. The speed of sound depends on the type of gas (for example air, pure oxygen, helium, carbon dioxide, etc.) and the absolute temperature of the gas.

The speed of sound in air depends on the type of gas and its temperature. On Earth, the atmosphere consists mostly of diatomic nitrogen (N_2) and diatomic oxygen (O_2), and the temperature has a rather complex relationship with altitude. (See next chapter.) Atmospheric scientists and aeronautical engineers have created empirical models of Earth's atmosphere to account for the variation of temperature with altitude and to include, when appropriate, the influence of trace gases such as carbon dioxide and water vapor.

As an aircraft moves through the air, it creates pressure waves. These pressure waves propagate away from the aircraft at the speed of sound. As the aircraft approaches the speed of sound (a condition called Mach1, discussed shortly), the aircraft begins to catch up to its own pressure waves. These pressure wave begin to build up at the front of the aircraft and turn into one large shock wave. This shock wave begins to buffet the aircraft and represents the so-called sound barrier. The shock wave also produces high drag (resistance to forward motion through a fluid) on the aircraft. As the aircraft passes through the shock wave, it is moving faster than the sound (pressure waves) it is creating. The shock wave forms an invisible cone of sound called the Mach cone that stretches out toward the ground. When the shock wave hits the ground, it causes a sonic boom.

The ratio of the speed of an object (such as an aircraft) relative to the speed of sound in a gas (such as air) determines the magnitude of many of the compressibility effects. Because of the importance of this ratio, scientists and engineers designated this special parameter as the Mach number (M). This designation honors the 19th century Austrian physicist Ernest Mach, who investigated the shock waves created by bullets and artillery shells as they passed through air at sonic and supersonic speeds. Contrary to his important contributions to gas dynamics, Mach steadfastly rejected the reality of atoms and the notion of kinetic theory until his death in 1916.

The Mach number (M) is a dimensionless number that expresses the ratio of the speed of a body (or of a point on that body) to the speed of sound in the medium through which the body is traveling. The Mach number can also refer to the speed of gas flow

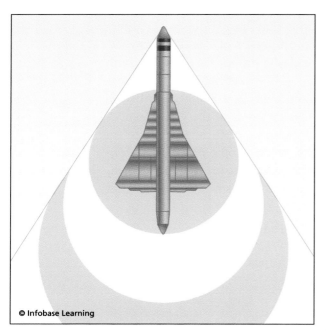

© Infobase Learning

This illustration shows a jet aircraft flying at supersonic speed (M > 1) with shock waves moving away from and behind the aircraft. When the shock waves reach the ground, they cause a sonic boom. *(NASA)*

Water vapor builds up around a U.S. Navy F/A-18 Super Hornet as it breaks the sound barrier during a flyby near the aircraft carrier USS *Kitty Hawk* (CV 63), which was operating in the Philippine Sea on July 27, 2005. *(USN)*

through an object, such as a nozzle or wind tunnel. If the Mach number is less than unity (that is, M < 1), scientists call the flow (or object moving through a gas) *subsonic,* and local disturbances can propagate ahead of the flow (or object moving through a gas). If the Mach number equals unity (i.e., M = 1), the object is traveling at sonic speed through the gas, or the flow is considered sonic. Aerospace engineers often call the Mach number range from M = 0.7 to M = 0.9 the transonic region. If the Mach number is greater than unity (i.e., M > 1), scientists regard the flow as supersonic or say that the object is traveling at supersonic speed through the gas. Under supersonic conditions, disturbances cannot propagate ahead of the object or the flow. Therefore, oblique shock waves form. (An oblique shock wave is inclined in the direction of flow.) The formation and characteristics of oblique shock waves involve complicated compressible flow phenomena.

A sonic boom is a noise, resembling thunder, caused by a shock wave that emanates from an aircraft, projectile, or other flying object, which is traveling at or above sonic velocity in Earth's atmosphere. As supersonic objects travel through the air, the air molecules are pushed aside

SHOCK WAVE

Physicists use the term *shock wave* in two different but complementary ways. In the first meaning, a shock wave is a surface of discontinuity (that is, an abrupt change of conditions) set up in a supersonic field of flow through which a fluid undergoes a finite decrease in velocity accompanied by a marked increase in pressure, density, temperature, and entropy. In the second meaning, a shock wave is the pressure pulse in air, water, or earth that propagates from an explosion. The explosion-related shock wave has two distinct phases. In the first (or positive) phase, the pressure rises sharply to a peak and then subsides to the normal pressure of the surrounding medium. In

The 15-kiloton (kT) yield nuclear explosion (code-named Grable) detonated at the Nevada Test Site on May 25, 1953 *(DOE/NTS)*

the second (or negative) phase, the pressure falls below that of the medium and then returns to normal. Scientists usually call an explosion-related shock wave that propagates in air a blast wave. Some 40 to 60 percent of the energy released in an atmospheric nuclear detonation appears as the blast wave.

with great force, and this action forms a shock wave. The bigger and more massive the supersonic object, the more air is displaced and the stronger the resultant shock waves. Several factors can influence sonic booms: the mass, size, and shape of the supersonic vehicle or object; its altitude, attitude, and flight path; and local weather and atmospheric conditions.

The shock wave forms a cone of pressurized air molecules that moves outward and rearward in all directions and extends to the ground. As the cone spreads across the landscape along the flight path, it creates a continuous sonic boom the full width of its base.

PRANDTL AND BOUNDARY LAYER THEORY

Scientists define a *boundary layer* as the layer of fluid in the immediate vicinity of a bounding surface. The German physicist Ludwig Prandtl (1875–1953) revolutionized fluid science when he presented his boundary layer concept in 1904. His historic publication was entitled "On the Motion of Fluids of Very Small Viscosity." He discovered that while the bulk of a flowing fluid could be adequately treated using classical potential flow techniques, there was a thin region near an object where viscous effects dominated. Prandtl's work led to a much better understanding of skin friction, a contribution of great importance to the emerging aeronautics industry.

In fluid science, the boundary layer corresponds to the layer of fluid affected by the viscosity of the fluid. Within this thin layer of fluid near

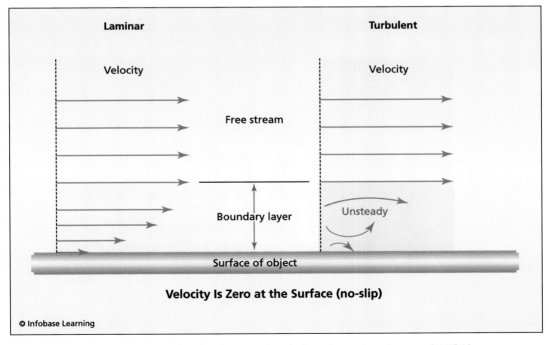

A simplified diagram comparing laminar and turbulent boundary layers *(NASA)*

the surface, the fluid velocity changes from its free stream velocity value to zero (at the surface or wall). Engineers often measure the thickness of the boundary layer from the surface to the point at which the fluid flow has 99 percent of its free stream velocity. Boundary layers can be either laminar (layered flow) or turbulent (disordered), depending upon the Reynolds number.

The Irish-born British engineer Osborne Reynolds (1842–1912) investigated the flow of fluids, especially in pipes. He developed an important dimensionless ratio, now called the Reynolds number, to characterize a fluid's dynamic state. The Reynolds number (Re) predicts changes in fluid flow regimes (that is, laminar flow, transition flow, or turbulent flow). It is defined as the ratio of inertia force (momentum force) to viscous force in fluid flow. Engineers often use the following mathematical expression for the Reynolds number: $Re = (\rho\ l\ v)/\mu$, where ρ is the fluid density, l is the characteristic length of the fluid system, v is the fluid velocity, and μ is the absolute viscosity. At low Reynolds number, the viscous force dominates, and the fluid flow and the motion is laminar. At high Reynolds number, the inertia force dominates, leading to a condition of turbulent flow.

The Prandtl number (Pr) is an important nondimensional parameter that scientists and engineers use in many heat transfer and mass transfer problems involving fluid flow. In mass transfer problems, the Prandtl number expresses the ratio of momentum diffusivity to mass diffusivity. Expressed mathematically, this ratio is $Pr = \mu/(\rho\ D)$, where μ is the absolute viscosity, ρ is the density, and D is the mass diffusivity.

CHAPTER 5

Earth's Atmosphere

This chapter discusses the physical properties and characteristics of Earth's atmosphere. Scientists define *atmosphere* as the gravitationally bound gaseous envelope that forms the outer region around a planet or other celestial body. Earth's atmosphere is a remarkable paradox. On one hand, the thin gaseous envelope contains life-sustaining gases and protects the creatures that inhabit the biosphere from space radiation and rogue meteoroids; on the other hand, the atmosphere gives rise to powerful, life-endangering storms such as hurricanes and tornadoes. Earth's atmosphere plays a major role in the hydrologic cycle and in the planet's overall energy balance. The winds, clouds, and air pollution are mentioned briefly here and then treated more extensively in subsequent chapters.

PHYSICAL PROPERTIES AND STRUCTURE OF EARTH'S ATMOSPHERE

When viewed edge-on from space, Earth's atmosphere appears as a very thin, gaseous envelope around humans' home planet. However, the atmosphere actually extends from Earth's surface out many thousands of miles (km). While the atmosphere becomes increasingly thinner (less dense) with height, its constituent molecules remain influenced by Earth's gravitational pull. The atmosphere is a mixture of life-sustaining gases and tiny

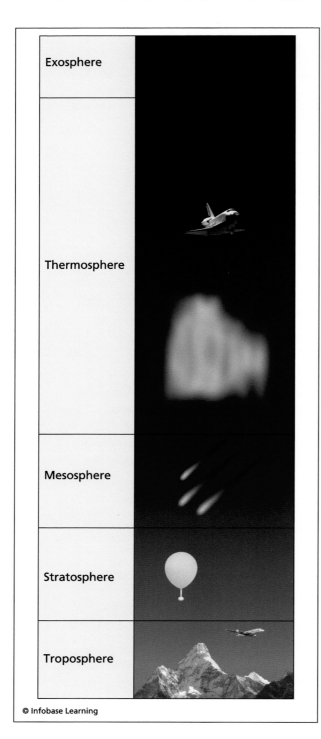

Exosphere

Thermosphere

Mesosphere

Stratosphere

Troposphere

© Infobase Learning

suspended particles called *aerosols.* Aerosols arise from a variety of sources such as volcanic eruptions, soil erosion, soot from fires, sea spray, and airborne pollutants from human activities.

Near sea level, Earth's atmosphere contains the following dry composition of gases (by volume): nitrogen (N_2), 78.084 percent; oxygen (O_2), 20.947 percent; argon (Ar), 0.934 percent; and carbon dioxide (CO_2), 0.033 percent. These four components make up 99.998 percent of all gases in the life-sustaining mixture people commonly refer to as air. Nitrogen is by far the most common gas in the atmosphere. Its presence dilutes oxygen and prevents rapid burning at Earth's surface. Living things need nitrogen to produce proteins. Oxygen is used by all living things and is essential for respiration (breathing). The oxygen in Earth's atmosphere is also necessary for combustion, or burning. Argon is an inert noble gas. Plants require carbon dioxide to manufacture oxygen during photosynthesis. Carbon dioxide is a greenhouse gas

This illustration shows the five basic layers of Earth's atmosphere. The vast majority (some 99 percent) of the air in Earth's atmosphere is located in the first two layers, called the troposphere and stratosphere, respectively. *(NOAA)*

AN ASTRONAUT'S VIEW OF EARTH'S ATMOSPHERE

The accompanying picture is a spectacular view of sunset on the Indian Ocean captured by astronauts aboard the *International Space Station* in May 2010. The image provides the limb (edge-on) view of Earth's atmosphere as observed from an Earth-orbiting spacecraft. The curvature of Earth is visible along the limb (horizon) line, which extends across the image from the center left side to the lower right side. Above Earth's darkened surface, there is a brilliant sequence of colors that roughly denote several layers of the atmosphere. Deep oranges and yellows appear in the troposphere. The troposphere is the lowest layer of the atmosphere and contains more than 80 percent of its mass, including almost all the water vapor, clouds, and precipitation. Variations in the colors are due primarily to varying concentrations of either clouds or aerosols. Several dark clouds are also visible within the troposphere.

The stratosphere appears as a thin pink to white region just above the layer of darkened clouds in the troposphere. Generally, this layer contains few, if any, clouds. The blue layers that appear above the stratosphere mark the transition between the middle and upper atmosphere. Finally, the image depicts how the upper atmosphere gradually fades into the blackness of outer space.

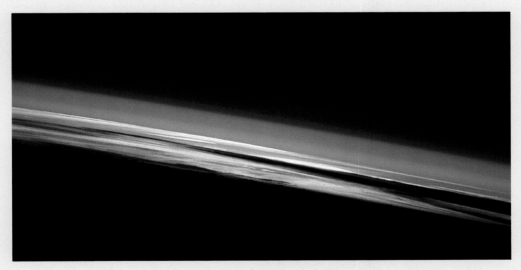

Astronauts aboard the *International Space Station* took this spectacular image on May 25, 2010, of sunset on the Indian Ocean. It provides a limb (edge-on) view of Earth's atmosphere as seen from orbit. *(NASA)*

that acts as a blanket to prevent heat from escaping into outer space. There are also lesser amounts of many other gases, including water vapor (H_2O) and human-generated chemical pollutants, such as chloro-fluorocarbons (CFCs).

Based on contemporary data from the National Oceanic and Atmospheric Administration (NOAA), some of the other gases found mixed in Earth's lower atmosphere include neon (Ne) at 18.20 parts per million (ppm); helium (He) at 5.20 ppm; krypton (Kr) at 1.10 ppm; sulfur dioxide (SO_2) at 1.0 ppm; methane (CH_4) at 2.0 ppm; hydrogen (H_2) at 0.50 ppm; nitrous oxide (N_2O) at 0.5 ppm; xenon (Xe) at 0.09 ppm; ozone (O_3) at 0.07 ppm; nitrogen dioxide (NO_2) at 0.02 ppm; and iodine (I_2) at 0.01 ppm. Scientists have also measured trace quantities of carbon monoxide (CO) and ammonia (NH_3) as part of the atmosphere's normal overall dry composition.

Meteorologists recognize that Earth's atmosphere is rarely, if ever, completely dry. Water vapor (H_2O) is nearly always present—at times up to a maximum quantity of 4 percent by volume. When hot, dry winds are blowing in desert regions, the atmosphere's water vapor content approaches zero. On very warm, humid days in subtropical and temperate climate regions, the water vapor content of the atmosphere often reaches about 3 percent by volume. The air in tropical climates reaches the upper limit of atmospheric water vapor content, approximately 4 percent by volume. The composition of such very "wet" tropical air is as follows: nitrogen, 74.96 percent by volume; oxygen, 20.11 percent; water vapor, 4.0 percent; argon, 0.89 percent; and 0.04 percent for all other gases combined.

Scientists use the general term *humidity* to describe the wetness of the atmosphere—that is, the amount of water vapor in the air for specific temperature and pressure conditions. The relative humidity represents the ratio of the amount of water vapor present in the air at a given temperature and atmospheric pressure to the greatest amount of water vapor content possible for those same conditions of temperature and pressure. A 10 percent relative humidity represents very dry air, while a 90 percent relative humidity represents very moist air.

Using various physical characteristics, such as how temperature and pressure change with altitude, chemical composition, and density, scientists have divided Earth's atmosphere into five distinct layers: the troposphere, stratosphere, mesosphere, thermosphere, and exosphere.

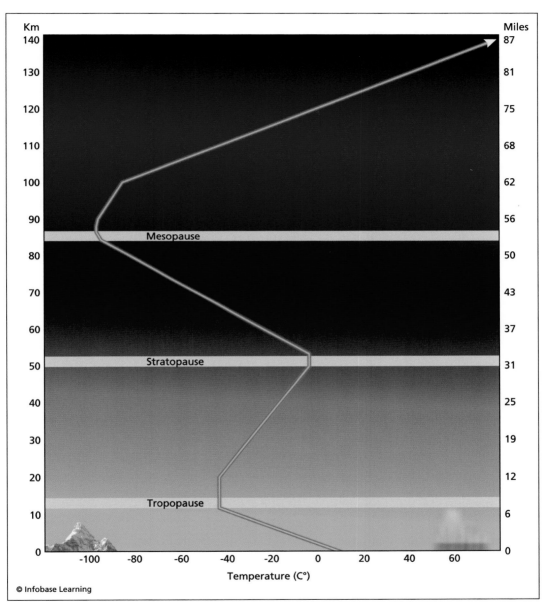

© Infobase Learning

The diagram depicts the average temperature profile through the lower layers of Earth's atmosphere. Altitude (in kilometers and miles) is indicated along the vertical axis; temperature (in degrees Celsius) appears along the horizontal axis. NOAA scientists point out that temperatures in the thermosphere continue to climb with altitude, reaching as high as 3,632°F (2,000°C). *(NOAA)*

The troposphere starts at Earth's surface and reaches four to 12 miles (6.4 to 25.8 km) altitude, depending on geographic location. (Scientists and aviators define *altitude* as the vertical distance, or height, above mean sea level.) At Earth's poles, the troposphere has a height of about four miles (6.4 km); at the equator, it reaches an altitude of about 12 miles (25.8 km). The troposphere is the warmest layer of the atmosphere and the part where life naturally occurs. Thanks to modern aerospace technology, human beings can now travel comfortably through the stratosphere, and a select few people (called astronauts, cosmonauts, or taikonauts) can even venture into the thermosphere and beyond. (Chapter 8 discusses human flight through the atmosphere.) Most of

THE IONOSPHERE

Located mostly within the thermosphere, the ionosphere consists of electrically charged gas particles that have become ionized as a result of the absorption of solar radiation, especially ultraviolet and X-rays. Depending on solar activity, the ionosphere can extend from an altitude of 37 to more than 600 miles (60–1,000 km). Scientists have divided the ionosphere into three layers: the F-layer (outermost), the E-layer (middle), and the D-layer (innermost). During the daytime, the F-layer splits into two layers, which recombine at night.

In December 1901, the Italian physicist Guglielmo Marconi (1874–1937) transmitted a wireless signal across the Atlantic Ocean from Cornwall, Great Britain, to Newfoundland, Canada. Immediately following Marconi's technical demonstration, the British physicist Oliver Heaviside (1850–1925) and the Irish-American electrical engineer Arthur Edwin Kennelly (1861–1939) independently postulated the existence of an ionized layer in Earth's upper atmosphere capable of reflecting radio waves. In the 1920s, the British physicist Sir Edward Victor Appleton (1892–1965) and other scientists experimentally demonstrated the existence of the ionosphere and began mapping its different layers. Because of the ionosphere, wireless radio wave messages can be transmitted over great distances despite Earth's curvature.

The ionosphere's density changes from day to night, with all three layers being denser during the day. At night, the D-layer decreases in ionized particle density, often to the point that it essentially disappears.

Earth's weather occurs in the troposphere. Thermal instability in the troposphere is largely responsible for the global patterns of atmospheric circulation. (Chapter 6 discusses the winds and global circulation patterns.) Scientists use the term *tropopause* to describe the transition boundary between the troposphere and the layer above it, that is, the stratosphere. They often regard the troposphere and the tropopause as the lower atmosphere.

The stratosphere extends from the tropopause up to an altitude of 31 miles (50 km). This layer contains 19 percent of the atmosphere's gases but has very little water vapor. Temperatures increase with height due to the increasing absorption of solar radiation by oxygen molecules, leading to the formation of ozone. As shown in the accompanying diagram, atmospheric temperature rises from an average value of –76°F (–60°C) at the tropopause to a maximum value of approximately 5°F (–15°C) at the stratopause due to the absorption of ultraviolet radiation from the Sun. The vast majority (some 99 percent) of the air in Earth's atmosphere is located within these two layers.

The stratopause separates the stratosphere from the next layer of the atmosphere, called the *mesosphere*. Scientists define Earth's middle atmosphere as the stratosphere and mesosphere, along with their respective transition regions, namely, the stratopause and mesopause. The mesosphere extends from the stratopause to an altitude of about 53 miles (85 km). With rising altitude, the atmosphere becomes thinner and thinner. Nevertheless, there are enough gas molecules in the mesosphere to slow down and burn up (by aerodynamic friction) the numerous small meteoroids that arrive from outer space.

Astronomers use the term *meteor* for the meteoroids that burn up in the atmosphere and appear in the night sky as flashes of light. They use the term *meteorite* to describe any space rock (less than 330 feet [100 m] in diameter) that survives its fiery plunge through Earth's protective atmosphere and lands on the planet's surface. Impacting space rocks with diameters larger than 330 feet (100 m) are called asteroids.

The temperature in the mesosphere decreases with increasing altitude. On average, the temperature at the stratopause is 5°F (–15C°). It falls to –184 °F (–120C°) at the mesopause.

The thermosphere extends from the mesopause to an altitude of about 430 miles (692 km). Scientists regard this layer as Earth's upper atmosphere. Human-crewed, Earth-orbiting spacecraft typically operate

SOLAR WIND

The solar wind is the variable stream of electrons, protons, and various atomic nuclei (such as alpha particles) that flows continuously outward from the Sun into interplanetary space. The source of the solar wind is the Sun's corona. The solar wind has a typical speed of about 250 miles per second (400 km/s). Although it is always directed outward from the Sun in all directions, the speed of the solar wind is not uniform. When the solar wind reaches Earth, only a small percentage of its energy penetrates the planet's magnetosphere, but the energy that does penetrate the planet's protective magnetic bubble is sufficient to create the brilliant aurora seen around Earth's poles and to endanger many high-technology electronic systems. The auroras are the visible glow in Earth's ionosphere produced by the interaction of the solar wind with the planet's magnetosphere. Energetic charged particles in the solar wind tend to follow Earth's magnetic field lines down toward the polar regions. Space scientists now recognize that the solar wind and Earth's magnetosphere influence each, since changes in the solar wind are capable of distorting Earth's magnetosphere.

Artist's rendering of the solar wind interacting with Earth's magnetosphere (not drawn to scale) *(NASA)*

in the upper portions of the thermosphere. As altitude increases, the gases of the thermosphere become increasingly thinner than in the mesosphere. The high-energy components of incoming solar radiation, namely, ultraviolet and X-ray photons, are absorbed by the gas molecules in the thermosphere. As the gas molecules absorb energetic solar photons, their kinetic energies greatly increase, causing the gas temperature to undergo a large increase. Near the thermopause, the apparent gas temperature (based on average molecular velocities) can be as high as 3,632°F (2,000°C). Scientists point out, however, that because the air is so thin at these altitudes, there are only a very few high-velocity molecules per unit volume, so the thermosphere is still very cold from the perspective of sensible heat.

In the thermosphere, temperature rises with altitude as oxygen and nitrogen molecules absorb ultraviolet radiation and X-rays from the Sun. This heating process also creates ionized atoms and molecules that form the special region called the *ionosphere,* which is characterized by the presence of electrically charged gases, or plasma. The boundaries and structure of the ionosphere vary considerably according to solar activity.

Scientists regard the exosphere as the outermost layer of Earth's atmosphere. The layer extends from the thermopause to a nominal altitude of 6,200 miles (10,000 km). In this layer of the atmosphere, collisions with energetic particles in the solar wind and very energetic solar photons allow individual gas molecules and atoms to reach sufficiently high velocities to escape Earth's gravitation pull. The escape velocity for Earth is approximately 7.0 miles per second (11.2 km/s). For comparison, the escape velocity from Mars is 3.1 miles per second (5.0 km/s).

PRESSURE VARIATION WITH ALTITUDE

Gravity is responsible for holding the molecules and atoms that make up Earth's atmosphere next to the planet's surface. Because of the attraction of gravity, the gas molecules near the surface are squeezed together by the weight of the gases above, giving rise to the phenomenon of atmospheric pressure. Starting in the 17th century, scientists such as Torricelli and Pascal began examining atmospheric pressure and soon discovered that it declined quite dramatically with increasing altitude.

The accompanying diagram shows how atmospheric pressure, expressed in millibars (mb) on the left vertical, varies with altitude,

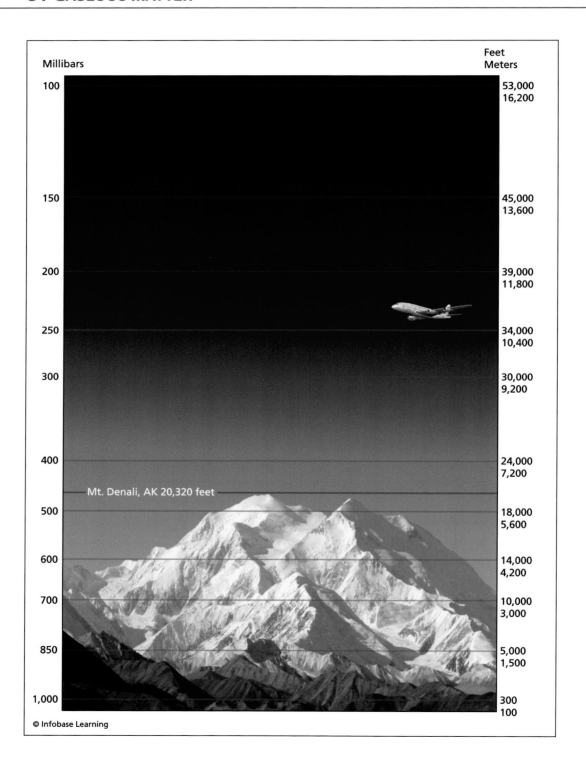

Millibars

Feet
Meters

100 — 53,000 / 16,200

150 — 45,000 / 13,600

200 — 39,000 / 11,800

250 — 34,000 / 10,400

300 — 30,000 / 9,200

400 — 24,000 / 7,200

Mt. Denali, AK 20,320 feet

500 — 18,000 / 5,600

600 — 14,000 / 4,200

700 — 10,000 / 3,000

850 — 5,000 / 1,500

1,000 — 300 / 100

© Infobase Learning

expressed in feet (ft) and meters (m) on the right vertical axis. The central portion of the illustration depicts Mount Denali (formerly Mount McKinley), Alaska—the highest peak in North America. The figure also shows the typical modern commercial jet airliner at its nominal cruising altitude. Atmospheric pressure decreases quite rapidly with altitude. In fact, half of all the air molecules in the atmosphere are contained within the first 18,000 feet (5,500 m) of altitude.

Because of this decrease of pressure with height, scientists found it quite difficult to compare the air pressure at one location on Earth with that measured at another location, especially when the individual research sites were situated at different elevations above sea level. As Torricelli first observed in 1643 with his newly invented mercury barometer, there are also important daily variations in atmospheric pressure at the same location due to changing weather conditions. The solution scientists employed was to convert the air pressure readings they collected at various stations to a value with an agreed upon common denominator. The common denominator they selected was sea level. Today, at meteorological observation sites around the world (regardless of a particular station's actual elevation), scientists use a series of calculations to convert a measured value of atmospheric pressure to the value that the instrument would have observed if that instrument were actually located at sea level.

The two most common units of atmospheric pressure used by meteorologists in the United States are inches of mercury (in Hg) and millibars (mb). By international agreement, the standard atmosphere at sea level exerts a force of 101,325 newtons per square meter (N/m^2 or pascals)—a value corresponding to a barometric reading of 29.92 inches (760 mm) of mercury.

The terms *bar* and *millibar* entered the lexicon of science in the early part of the 20th century due to the work of the British meteorologist Sir William Napier Shaw (1854–1945). The word *bar* derives from the ancient Greek word *baros* (βαρος), meaning weight. In 1929, the National Weather Service of the United States adopted the millibar (mb) and defined it as

---

(opposite page) This chart shows the average heights of pressure in the atmosphere. Denali (formerly Mount McKinley) in Alaska is the highest peak in North America. Commercial jet aircraft typically cruise at an altitude of about 35,000 feet (10,670 m), where the average atmospheric pressure is 240 millibars. *(NOAA)*

being equal to a pressure of 100 N/m², or 100 pascals. (One millibar is one-thousandth (1/1,000) of a bar.)

Since then, American meteorologists have reported the standard pressure at sea level as approximately one bar (or 1,000 mb), rather than the more exact value of 101,325 pascals. Due to tradition and convenience, meteorologists generally ignore the relatively small difference in pressure values at sea level.

The range of millibar values meteorologists encounter in weather forecasting extends from about 100 to 1,050 mb. On weather maps, they express the atmospheric pressure at Earth's surface (referred to as mean sea level, or MSL) in terms of millibars. In addition to sea-level measurements, weather forecasters generally favor examining atmospheric conditions at the 500 mb level. This pressure level corresponds to an altitude of about 18,000 feet (5,500 m), a good height at which to observe the movement of transient waves in the atmosphere.

Personnel at the National Oceanic and Atmospheric Administration (NOAA) and other agencies of the federal government use a series of models called the U.S. Standard Atmosphere that define values for atmospheric properties such as pressure, temperature, density over a wide range of altitudes. In the current (1976) version of the U.S. Standard Atmosphere, the static atmospheric pressure at sea level is expressed as 101,325 pascals, 29.92126 inches (760 mm) of mercury (Hg), or 14.696 lbf/in² (psi). The corresponding standard atmospheric temperature is expressed as 288.15 K, a value of absolute temperature that represents a relative temperature of 59°F (15°C). The mean sea level (MSL) value of density for a dry, perfect mixture of air is 0.0764734 lbm/ft³ (1.225 kg/m³).

Chemists and physicists often compare volumes of gases at standard temperature and pressure (STP). By convention within the international scientific community, the reference conditions for gases is 0°C (32°F) and one atmosphere (1 atm) pressure. The standard atmosphere has a pressure of 101,325 pascals or 29.92 inches (760 mm) of Hg. The standard atmosphere is commonly equated to an atmospheric pressure value of one bar, although one bar was originally defined as representing 100,000 pascals.

EARTH SYSTEM SCIENCE

From the vantage point of outer space, scientists and their automated spacecraft can view the Earth as a whole system, observe the results of complex interactions, and begin to understand how the planet is changing. They are using the unique view from space to study the Earth as an integrated system. Scientists define *Earth system science* (ESS) as the contemporary study of planet Earth facilitated by space-based observations and sophisticated computer-based climate models.

Earth system science is a multidisciplined scientific endeavor that treats humans' home planet as an interactive, complex system. The five major components (or interlinked "spheres") of the Earth system are the atmosphere, the hydrosphere (primarily the oceans), the cryosphere (ice and snow), the biosphere (which includes all living things), and the solid Earth (especially the planet's surface and soil). Scientists and

GAIA HYPOTHESIS

The Gaia hypothesis is the interesting but highly speculative idea first suggested in 1969 by the British independent scientist James Lovelock (1919–   ) with the assistance of the American biologist Lynn Margulis (1938–   ) that Earth's biosphere has an important modulating effect on the terrestrial atmosphere. Because of the chemical complexity observed in the lower atmosphere, Lovelock has suggested that life-forms within the terrestrial biosphere actually help control the chemical composition of the Earth's atmosphere, thereby ensuring the continuation of conditions suitable for life. Gas-exchanging microorganisms, for example, are thought to play a key role in this continuous process of environmental regulation. Without these cooperative interactions, in which some organisms generate certain gases and carbon compounds that are subsequently removed and used by other organisms, planet Earth might possess an excessively hot or cold planetary surface devoid of liquid water and surrounded by an inanimate, carbon dioxide–rich atmosphere.

Gaia (also spelled Gaea) was the goddess of Earth in ancient Greek mythology. At the suggestion of the British novelist William Golding (1911–93), Lovelock used her name to represent the terrestrial biosphere—namely, the system of life on Earth, including living organisms and their required liquids, gases, and solids. Thus, the Gaia hypothesis simply states that Gaia (a mythology-based personification of Earth's biosphere) will struggle to maintain the atmospheric conditions suitable for the survival of terrestrial life.

nonscientists alike have found it helpful to personify the concept of an integrated Earth system as *Gaia*—the Mother Earth goddess found in Greco-Roman mythology.

As they learn more about the linkages of complex environmental processes, scientists are developing an improved prediction capability for climate, weather, and natural hazards. Today, the enormous streams of data from Earth-observing satellites are promoting collaboration by scientists from many different disciplines within a common multifaceted discipline known as Earth system science (ESS). Through the analysis of data derived from modern spacecraft, scientists are acquiring an improved understanding of Earth as a complex, highly interactive system.

Scientists consider models an essential tool for combining observations, theory, and experimental results. Through the use of advanced models, they can investigate how the Earth system functions and how the complex system is affected by natural phenomena as well as human activities. Comprehensive climate models generally include the major components of Earth's climate system (that is, the atmosphere, hydrosphere, land surface, cryosphere, and biosphere) and can account for the transfer of energy, water, mass, and organic chemicals among these components. Although scientists have enjoyed significant progress in the past decade, their climate modeling efforts remain works in progress.

Of special importance here is the fact that the atmosphere links all the major components of the Earth system. The atmosphere interacts with the oceans (hydrosphere), land, terrestrial and marine plants and animals (biosphere), and frozen regions (cryosphere). As a result of these linkages, the atmosphere serves as a conduit of change. Emissions from natural sources and human activities enter the atmosphere at the surface and are then transported to other geographical locations, including those at higher altitudes. Some emissions experience chemical transformation or removal while in the atmosphere or else interact with the process of cloud formation and precipitation. Some natural events such as volcanic eruptions and large-scale human activities such as the release of carbon dioxide due to the combustion of fossil fuels can change the atmosphere's composition, thereby influencing the Earth's radiation (energy) balance (ERB).

Another major feature of the atmosphere is that is serves as a long-term reservoir for certain trace gases such as carbon dioxide, which can

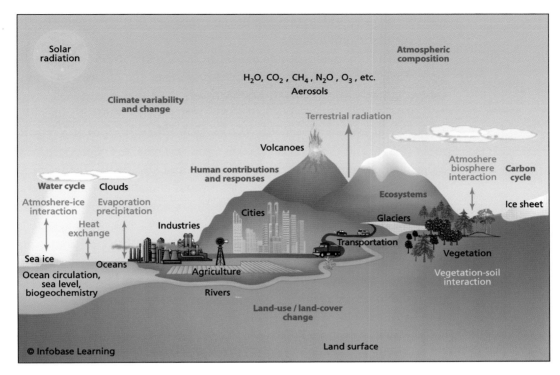

This diagram shows the major components needed to understand Earth's climate system and the process of climate change. Scientists are beginning to appreciate how complex the coupling between the atmosphere, hydrosphere (oceans), cryosphere (ice), and biosphere really is. They use models of varying complexity to forecast and project climate phenomena. *(Adapted from the International Panel on Climate Change [IPCC])*

cause global climate changes. Scientists monitor for atmospheric composition changes that would indicate the onset of adverse environmental conditions that would affect all countries and populations. Because of the very long removal times of some gases from the atmosphere (for example, more than 100 years for carbon dioxide), any global changes associated with such trace gases could persist for decades, centuries, or even millennia.

The climate system is dynamic, and modeling is the only way scientists can effectively integrate the current knowledge of its individual components. Through modeling studies, they can estimate and project the future state of the climate system. At present, however, they do not have a complete understanding of the processes that contribute to climate variability and change.

Scientists are investigating changes in the Earth's atmospheric chemistry. Their research involves the following areas: the changes in atmospheric composition and the timescales over which they occur, the forcing functions (human-generated and natural) that drive these changes, the reactions of trace components in the atmosphere to global environment change and the resulting effects on the climate, the effects of global atmospheric chemical and climate changes, and air quality. The relationship between Earth's atmosphere and ground emissions presents several important environmental issues, including global ozone depletion and global air quality.

The first weather satellite expanded the possibilities of predicting tomorrow's weather. With advanced spacecraft, comprehensive data gathering, and improved modeling, meteorologists can refine their predictive capabilities. Accurately predicting changes in ozone, air quality, and climate helps scientists and nonscientists alike understand how human activities are impacting Earth.

The weather system includes the dynamics of the atmosphere and its interaction with the oceans and land. Weather includes those local or microphysical processes that occur in minutes through the global-scale phenomena that can be predicted with a degree of success at an estimated maximum of two weeks ahead. Weather is an important part of Earth system science. An improvement in the scientific understanding of weather processes and phenomena is crucial to gaining a more accurate understanding of the overall Earth system. Weather is directly related to climate and to the water cycle and energy cycles that sustain life within the terrestrial biosphere.

The water cycle and the energy cycle involve the distribution, transport, and transformation of water and energy within the Earth system. Since solar energy drives the water cycle and energy exchanges are modulated by the interaction of water with radiation, the energy cycle and the water cycle are intimately entwined. Many issues remain to be resolved, including developing a precise understanding of the role of clouds. Of specific importance is the fact that clouds play a significant role on variability in the global water cycle and the planet's energy balance.

INTERGOVERNMENTAL PANEL ON CLIMATE CHANGE

The Intergovernmental Panel on Climate Change (IPCC) consists of government delegations from 194 countries as of July 2010. Membership is open to all member countries of the United Nations (UN) and the World Meteorological Organization (WMO). The IPCC is the leading body for the assessment of climate change. The organization was created in 1989 to provide the people of the planet a clear scientific view on the current state of climate change and its potential environmental and socioeconomic consequences. The IPCC is a scientific body whose members review and assess the most recent scientific, technical, and socioeconomic information produced worldwide that is relevant to the understanding of climate change. Although the IPCC does not conduct any research or environmental monitoring operations, thousands of scientists from around the world contribute to the work of the organization on a voluntary basis. Review is an essential part of the IPCC process, and IPCC reports often reflect differing viewpoints within the scientific community. The 2007 Nobel Peace Prize was jointly awarded to the IPCC and former U.S. vice president Albert Arnold (Al) Gore, Jr. (1948–   ), "for their efforts to build up and disseminate greater knowledge about man-made climate change, and to lay the foundations for the measures that are needed to counteract such change."

ATMOSPHERE-HYDROSPHERE INTERACTIONS

Earth's great hydrologic (or water) cycle consists of the continuous journey made by water molecules from the planet's surface, into the atmosphere, and back again. The hydrologic cycle has been operating for billions of years, and life on Earth depends upon it. Radiant energy from the Sun powers this planetary process, which involves the continuous exchange of moisture between the oceans, the atmosphere, and the land. As depicted in the illustration, there are many interesting components to Earth's water cycle. This section focuses on the following processes, which atmospheric scientists consider the most important: evaporation and transpiration, condensation, precipitation, and runoff and groundwater.

During the hydrologic cycle, radiant energy from the Sun causes large quantities of water to evaporate from the surface of the oceans and other

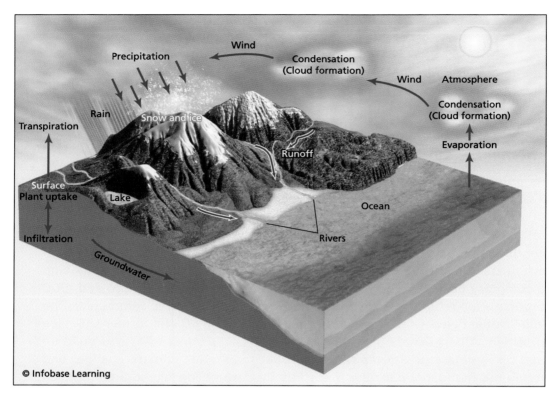

This figure shows the basic components of Earth's great hydrologic (water) cycle. *(NOAA)*

bodies of water, such as streams, rivers, and lakes, and enter the atmosphere. Water can also evaporate from moist soil. Scientists define *evaporation* as the physical process by which a liquid transforms into a gas (vapor) at a temperature below the boiling point of the liquid. Scientific studies indicate that the oceans, seas, and other bodies of surface water provide 90 percent of the moisture in Earth's atmosphere.

A little less than 10 percent of the moisture found in Earth's atmosphere is released by plants, trees, and other vegetation through the process of transpiration. Plants and trees collect water through their root systems as part of the delivery of nutrients to their leaves. They then release this water back into the atmosphere by transpiration, which is basically how plants sweat and cool themselves. In fluid mechanics, engineers describe *transpiration cooling* as a form of mass transfer that involves controlled injection of a fluid mass through a porous surface. This process is basically limited by the maximum rate at which the coolant material can be pumped through a porous surface.

Two other natural processes send water from Earth's surface back into the atmosphere. The first process involves sublimation from snow banks and ice fields. Scientists describe sublimation as the direct transition of a substance from the solid state (here, ice or snow) to the gaseous state (here, water vapor) without passing through the liquid state. Volcanic eruptions (see chapter 7) also inject water vapor and other gases into the atmosphere.

As air containing water vapor rises into the atmosphere, it expands and cools. When the ascending moist air cools to a certain temperature, called the dew point, it becomes a saturated vapor and begins to condense. Scientists define the *dew point* as the temperature at which water vapor condenses to liquid water when it is cooled at constant pressure. Clouds, fog, dew, mist, and frost are examples of condensation in the atmosphere.

Condensation is the change of state process by which a vapor (gas) becomes a liquid, the opposite of evaporation. During condensation in the atmosphere, moist air cools and loses its capacity to support water vapor. This excess water vapor subsequently condenses, forming droplets. Depending on the prevailing meteorological conditions, the newly formed droplets in a cloud can eventually grow and produce precipitation, including rain, snow, sleet, and hail. Precipitation is the primary mechanism within the hydrologic cycle by which water travels from the atmosphere back to Earth's surface.

When precipitation falls on an ocean or other large body of surface water, it gets stored and awaits conditions favorable for evaporation. When precipitation falls over land, water follows one of several routes. Some precipitation falls as snow and accumulates in ice caps or glaciers, which can store frozen water for thousands of years. As climate conditions change, the ice caps or glaciers can grow and store more frozen water, or they can shrink and retreat, melting in the process and returning the snowmelt runoff to streams and ultimately the oceans. An ice cap, glacier, or snowpack may also return a small quantity of moisture directly to the air through sublimation. In warmer, nonarctic climates, the annual thawing and melting of snowpacks releases the snowmelt— water that flows overland toward the oceans or large lakes. In many parts of the world, the snowpack represents an important water resource that annually feeds local streams and rivers as it melts. However, sudden melting of a snowpack due to unseasonably warm weather late in winter can cause severe flooding.

Generally, when precipitation falls on land, it will either be absorbed into the ground (eventually becoming groundwater), or, if the ground cannot absorb more water, the rainfall becomes runoff and flows into streams under the influence of gravity. Some of the groundwater is taken up by plants and trees and returns to the atmosphere through transpiration, while some of the runoff evaporates as it flows into streams or rivers. Part of the precipitation that infiltrates the ground eventually makes its way into streams. The water in streams then converges into rivers and flows back to the oceans, which contain some 97 percent of the planet's total supply of water. According to the scientists at the United States Geological Survey (USGS), the remaining 3 percent of the planet's water supply is freshwater. Of this amount of freshwater, about 69 percent is locked up in icecaps and glaciers, 30 percent is in the ground, and less than 1 percent resides in surface water sources. Finally, lakes contain about 87 percent of world's freshwater sources, swamps about 11 percent, and rivers about 2 percent.

EARTH'S RADIATION BUDGET

Scientists regard Earth's radiation budget (ERB), or overall energy balance, as the most fundamental physical quantity that influences the planet's climate. The ERB components include the incoming solar radiation; the solar radiation reflected back to space by the clouds, the atmosphere, and Earth's surface; and the long-wavelength thermal radiation emitted by Earth's surface and its atmosphere. The latitudinal variations of Earth's radiation budget are the ultimate driving force for the atmosphere and ocean circulations and hence the resulting planetary climate.

The Earth-atmosphere energy balance is a basic statement of the first law of thermodynamics, which describes the balance between incoming energy from the Sun and outgoing energy from Earth. Solar energy arrives at Earth as short-wavelength visible and ultraviolet radiation. When solar radiation reaches Earth, some is reflected directly back into space by clouds, some is absorbed by the atmosphere, and some is absorbed at the planet's surface (ocean or land). Since the Earth is much cooler than the Sun (288 K average surface temperature versus 5,770 K

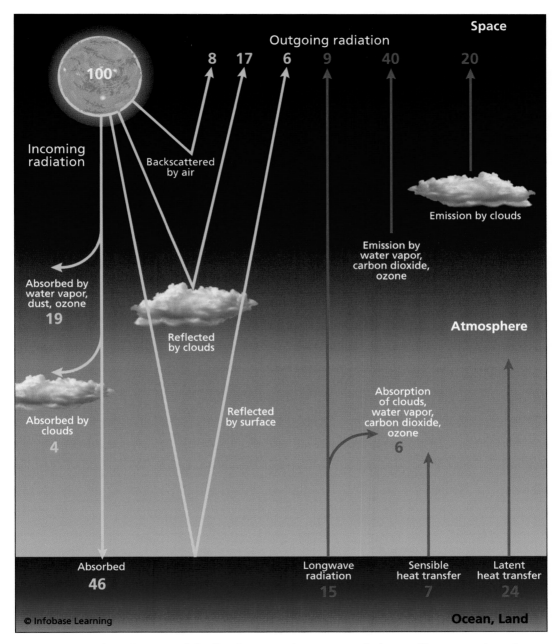

Space

Outgoing radiation

8   17   6   9   40   20

100

Incoming
radiation

Backscattered
by air

Absorbed by
water vapor,
dust, ozone
19

Emission by clouds

Emission by
water vapor,
carbon dioxide,
ozone

Reflected
by clouds

Atmosphere

Absorbed by
clouds
4

Reflected
by surface

Absorption
of clouds,
water vapor,
carbon dioxide,
ozone
6

Absorbed
46

Longwave
radiation
15

Sensible
heat transfer
7

Latent
heat transfer
24

© Infobase Learning

Ocean, Land

This is a simplified diagram of the Earth-atmosphere energy balance. Short-wavelength
solar energy (depicted in yellow as 100 arbitrary energy units) arrives at the top of Earth's
atmosphere, and energy (a total of 100 arbitrary units) leaves Earth as a combination of
short-wavelength and long-wavelength radiation. Thermal infrared (long-wavelength)
energy exchange quantities appear in red numerals. *(NOAA)*

surface temperature), it emits radiation (that is, radiates energy) at much longer thermal infrared wavelengths. Earth can only maintain a stable average temperature and therefore, a stable climate, if the total amount of energy arriving from the Sun in a unit time balances the amount of energy lost by Earth back to outer space in that same unit time.

The accompanying diagram depicts a simplified energy balance in terms of 100 arbitrary solar energy units (colored yellow), arriving at the top of Earth's atmosphere. Of these 100 arriving solar energy units, 31 units are immediately returned to space as short-wavelength radiation by air molecule scattering, cloud reflections, and planetary surface reflections. The rest of the incoming solar energy units are absorbed within the atmosphere or on the planet's surface. As part of this simplified overall energy balance, 69 energy units (at long-wavelength infrared energies depicted as red colored numerals) return to space. The illustration also shows the remainder of nominal energy exchanges within the atmosphere and between the atmosphere and the ocean or land.

SVANTE AUGUST ARRHENIUS

In 1895, the Swedish Nobel laureate and chemist Svante August Arrhenius (1859–1927) boldly ventured into the fields of climatology, geophysics, and planetary science when he presented an interesting paper to the Stockholm Physical Society. Entitled "On the Influence of Carbonic Acid in the Air upon the Temperature of the Ground," this visionary paper anticipated by decades contemporary concerns about the greenhouse effect and the rising carbon dioxide (carbonic acid) content in Earth's atmosphere. Arrhenius argued that variations in trace atmospheric constituents, especially carbon dioxide, could greatly influence Earth's overall heat (energy) budget.

During the next 10 years, Arrhenius continued his pioneering work on the effects of carbon dioxide on climate, including his concern about rising levels of anthropogenic (human-caused) carbon dioxide emissions. He summarized his major thoughts on the issue in the 1903 textbook *Lehrbuch der kosmichen Physik (Textbook of Cosmic Physics),* an interesting work that anticipated the development of several new scientific disciplines, including planetary science and Earth system science.

The absorption of longer-wavelength infrared radiation within the atmosphere as it tries to escape from Earth back to space is especially important to the global energy balance. Energy absorption within the atmosphere due to the greenhouse effect stores more energy near the planet's surface than would be stored if there were no greenhouse gases present. NASA planetary scientists note that if it were not for the naturally occurring greenhouse gases, Earth would have an average surface temperature of just –2°F (–18.9°C) rather than the 59°F (15°C) average surface temperature currently measured. All the surface water would be frozen, making the diversity of life in the biosphere impossible.

One of the most intriguing questions that faces atmospheric physicists and climate modelers is how clouds affect climate, and vice versa. Understanding these effects requires a detailed knowledge of how clouds absorb and reflect both incoming short-wavelength solar energy and outgoing long-wavelength (thermal infrared) terrestrial radiation. Contemporary data suggest that clouds that form over water are very different from clouds that form over land. These differences affect the way clouds reflect sunlight back into space and how much long-wavelength infrared energy from Earth the clouds absorb and re-emit.

Water vapor in the atmosphere also affects daily weather and long-term climate, because water vapor is a greenhouse gas that absorbs outgoing long-wavelength radiation from Earth. Because water vapor also condenses to form clouds, an increase in water vapor in the atmosphere may also increase the amount of clouds.

GREENHOUSE EFFECT

Scientists describe the greenhouse effect as the general warming of the lower layers of a planet's atmosphere due to the presence of gases such as water vapor (H_2O), carbon dioxide (CO_2), nitrous oxide (NO_2), and methane (CH_4). On Earth, the greenhouse effect takes place because the atmosphere is relatively transparent to visible light from the Sun (typically short-wavelength, 0.3 to 0.7 micrometer, radiation), but is essentially opaque to the longer-wavelength, nominally 10.6 micrometer, thermal infrared radiation emitted by the planet's surface. Because of the presence of greenhouse gases in Earth's atmosphere—including carbon dioxide, water vapor, methane, and human-made industrial gases such as hydrofluorocarbons (HFCs)—this outgoing thermal radiation

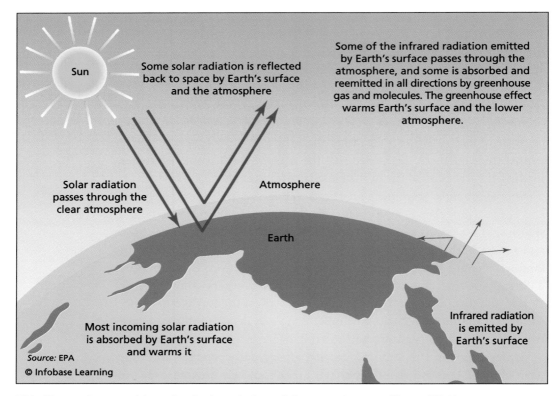

Sun

Some solar radiation is reflected back to space by Earth's surface and the atmosphere

Some of the infrared radiation emitted by Earth's surface passes through the atmosphere, and some is absorbed and reemitted in all directions by greenhouse gas and molecules. The greenhouse effect warms Earth's surface and the lower atmosphere.

Solar radiation passes through the clear atmosphere

Atmosphere

Earth

Most incoming solar radiation is absorbed by Earth's surface and warms it

Infrared radiation is emitted by Earth's surface

Source: EPA

© Infobase Learning

This illustration provides a basic description of the greenhouse effect. *(EPA)*

from Earth's surface is often blocked from escaping to space. The absorbed thermal energy can then produce a rise in the temperature of the lower atmosphere.

Climatologists suggest that the presence of greenhouse gases in Earth's early atmosphere made the planet habitable and prevented it from remaining a completely frozen wasteland. However, they also point out that as additional greenhouse gases accumulate in the atmosphere (primarily due to human activities), more outgoing thermal radiation could become trapped, producing a global warming trend.

Scientists around the world are concerned that the continued expansion of certain human activities, such as the burning of fossil fuels, will further increase the presence of greenhouse gases in the atmosphere and upset the planet's overall energy balance. Some of these scientists have suggested that human activities may be creating the conditions of a runaway greenhouse, as appears to have occurred in the past on the planet Venus. Such an effect is a planetary climatic extreme in which all of the surface water evaporates from the surface of a life-bearing or potentially life-bearing planet. Planetary scientists believe that the current atmosphere of Venus allows sunlight to reach the planet's surface, but its thick clouds and very rich carbon dioxide content (96.6 percent by volume) prevent surface heat from being radiated back to space. Such a condition has led to the evaporation of all surface water on Venus and has produced the present infernolike average surface temperature of approximately 873°F (467°C), a temperature hot enough to melt lead.

CHAPTER 6

Wind—Its Power and Applications

"It's hard to make predictions, especially about the future!"
—Yogi Berra, New York Yankees

This chapter describes the restless motions of air within the atmosphere known as the winds. The connection between global circulation patterns and weather is emphasized. The chapter also explains the origin and impact of severe weather phenomena, such a thunderstorms, tornados, and hurricanes, and how meteorologists now use atmospheric models to improve predictions and extend warning times. A brief summary of the role of wind power in human history completes the chapter.

GLOBAL CIRCULATION PATTERNS

The wind is simply air in motion. Winds blow between areas that possess different atmospheric pressures. Because Earth is spherical and not flat, incoming solar radiation does not fall evenly on the planet from the equator to the poles. The Sun shines more directly on regions around the equator, heating them more intensely. The result is that Earth's equatorial regions are warm all year long. In contrast, Earth's polar regions (the Arctic and Antarctic) are at such an oblique angle to the Sun that they experience significantly reduced amounts of incoming sunlight. The differences in atmospheric and surface temperatures from the equator to the poles set in motion a continuous movement of air and water around the planet. Great churning currents in the oceans and swirling winds in the atmosphere continuously carry heat from Earth's warmer equatorial regions up toward the colder polar regions.

Scientists define *weather* as the day-to-day result of Earth's churning atmosphere, which helps redistribute heat and water around the planet. They define *climate* as the average weather in a region over a period of many years. When scientists discuss climate, they are actually describing long-term (typically 30- to 50-year) patterns of weather (that is, temperature and precipitation) in particular regions of the world. People anticipate changes in weather from day to day, but the thought of significant changes in climate makes most people uneasy.

While discussing either weather or climate, scientists recognize that it is the atmosphere that links all the major components of the Earth system (atmosphere, hydrosphere, cryosphere, biosphere, and solid Earth). Because weather influences a person's daily activities, meteorologists

AIR MASSES

Atmospheric scientists define an air mass as a large body of air with a generally uniform value of temperature and humidity. They refer to the area from which an air mass originates as the source region. Based on the source region, there are four major types of air masses that influence weather across the continental United States. The categories of influential air masses are polar latitudes, continental, maritime, and tropical latitudes. For example, the Gulf Coast states and the eastern third of the country usually experience tropical air masses in summer.

Air masses can control weather for extended periods that range from days to months. Most weather events take place along the periphery of air masses at boundaries called fronts. Meteorologists commonly use such descriptive terms as cold front, warm front, and stationary front. Fronts are basically the boundaries between two different air masses. A front has vertical structure as well as spatial extent along Earth's surface. Cold fronts have steeper slopes that force air upward more abruptly. This condition usually creates a narrow band of showers and thunderstorms along or just ahead of the cold front.

attempt to make weather forecasts accurate out to several days, typically a week, but they face significant challenges when attempting to predict the occurrence and behavior of adverse weather phenomena such as thunderstorms, tornados, and hurricanes. Climatologists take a much longer view and construct complex models of the Earth system in their efforts to predict changes in climate for a particular region or the entire planet. Some scientists, called paleoclimatologists, look far back into Earth's past in efforts to develop better climate modeling tools.

The daily wind cycle plays a major role in establishing local and regional weather conditions. During the day, the air above the land heats up more quickly than the air over water. The warm air over land expands and rises, causing cooler air, which is denser or heavier, to rush in and take its place. At night, the conditions are reversed because the air cools more rapidly over land than over water. In a similar manner, the atmospheric winds that circle Earth are created because the surface near the planet's equator is heated more by the Sun than the surface near the North and South Poles. The actual process of wind formation is far more complicated.

CLIMATE CLASSIFICATIONS

While weather usually varies from day to day at any particular location on Earth, over long periods, such as years, the same type of weather will reoccur. Scientists call the recurring average weather found in any particular location that location's climate. The Russian-born German climatologist Wladimir Köppen (1846–1940) used the general temperature profile related to latitude to divide the world's climates into several major categories. These categories are briefly summarized here based on contemporary NOAA data. Tropical (moist) climates have average temperatures greater than 64°F (18°C) all months of the year and experience greater than 59 inches (150 cm) of precipitation annually. Dry climates are those in which potential evaporation and transpiration exceed precipitation. Moist subtropical midlatitude climates experience warm and humid summers along with mild winters. Moist continental midlatitude climates have warm to cool summers along with cold winters. Polar climates experience year-round cold temperatures, with the warmest month being less than 50°F (10°C). Finally, highlands possess unique climates based on their elevations. Highland climates occur in mountainous terrain, where rapid changes in elevation promote rapid climatic changes over relatively short distances.

Scientists use global circulations to explain how air masses and storm systems travel over Earth's surface. If Earth did not rotate, if the planet's axis of rotation was not tilted by 23.5° from the vertical, and if the planet did not possess large bodies of liquid water on its surface, global circulation would be quite simple and rather boring. In this physically bland hypothetical situation, the Sun would still heat the entire surface of the planet, but where the Sun appeared more directly overhead, more incoming solar energy would be available to heat the atmosphere and surface. The equatorial regions would become very hot, and the lower-density, hot air would rise into the upper atmosphere. The hot air would then move toward the poles, where it would become very cold and denser and sink toward the planet's surface. The colder air would then flow back to the equator and become available to repeat this relatively simple atmospheric circulation process. In this hypothetical planetary circulation model, Earth would have a large belt of low atmospheric pressure around the equator and a large area of high atmospheric pressure over each of the poles.

However, Earth rotates, has an axial tilt, and contains more land mass in the Northern Hemisphere than in the Southern Hemisphere. These factors and others make global circulation a far more complicated process. Instead of one large circulation pattern between the equator and the poles, for example, Earth's atmosphere actually has three major circulations: the Hadley cell, the Ferrel cell, and the polar cell, as noted in the accompanying illustration.

First proposed in 1735 by the British scientist George Hadley (1685–1768), the Hadley cell is a thermally driven atmospheric circulation cell that dominates tropical and subtropical climates. The Hadley cell involves low-latitude air movement toward the equator. When air masses in this cell are heated at the equator, they rise vertically into the upper atmosphere and move toward the poles. There is a Hadley cell in the Northern and Southern Hemispheres. Hadley suggested this type of circulation cell to partially explain the trade winds.

In the Ferrel cell, air near Earth's surface flows eastward and toward the North Pole in the Northern Hemisphere (and toward the South Pole in the Southern Hemisphere); air at higher altitudes flows westward and toward the equator. Scientists named this midlatitude atmospheric circulation cell after the American meteorologist William Ferrel (1817–91).

The polar cell dominates the atmosphere at high latitudes in both the Northern and Southern Hemispheres. In this circulation cell, air rises, diverges, and travels toward the poles. At the poles, the air becomes colder and sinks to form the polar highs. At the surface, the sinking air diverges outward away from the polar highs.

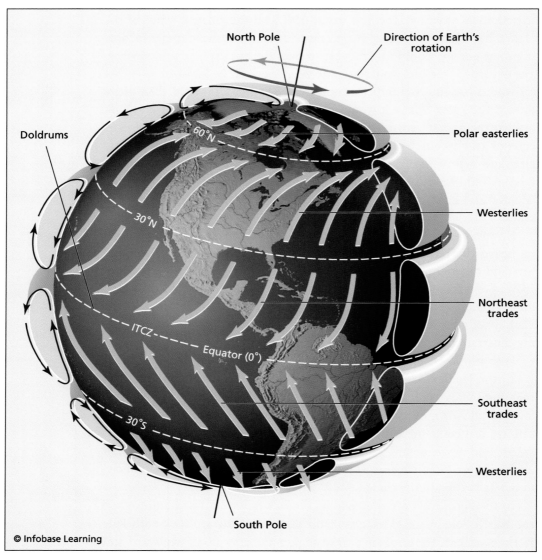

Global circulations explain how air and storm systems travel over Earth's surface. As shown in this illustration, instead of one large circulation between the poles and the equator, there are actually three circulations: low-latitude Hadley cell, mid-latitude Ferrel cell, and high-latitude polar cell. *(NOAA)*

Scientists describe the Coriolis effect as the apparent deflective effect on freely moving objects, including the atmosphere and oceans, caused by Earth's rotation. Atmospheric deflections attributed to the Coriolis effect move to the right in the Northern Hemisphere and to the left in the Southern Hemisphere. As it travels around Earth, the momentum of the air is conserved. This means that as air from the equator begins moving toward the North Pole, for example, it keeps its eastward motion constant. The surface of Earth below the moving air moves slower as the air passes overhead. The net result is that the air moves faster and faster in an easterly direction relative to the Earth's surface below the farther the air moves away from the equator toward the North Pole. (Remember, because of the dynamics of a rotating solid body, a person standing on the equator is actually moving much faster than a person standing on Earth's surface at the 45°N (or S) latitude line.) The phenomenon bears the name of the French physicist Gaspard de Coriolis (1792–1843), who introduced the notion in 1835. The Coriolis effect influences circulation patterns within Earth's atmosphere. This phenomenon generally accounts for the east-west direction of winds on Earth's surface.

In the zone between about 30°N and 30°S latitudes, the air at the surface flows toward the equator, and the air at higher altitude flows toward one of the poles. European mariners referred to the low-pressure area of calm, light, variable winds near the equator as the *doldrums.* In the subtropical high-pressure belts (at about 30°N latitude and 30°S latitudes), upper atmosphere air flowing toward the poles begins to cool and descend. This sinking air is relatively dry, because when it ascended into the atmosphere near the equator, the air released its moisture above the tropical rain forests. Winds at the surface are weak and variable near the center of this high-pressure zone of descending air, a region sometimes called the horse latitudes. (According to a rather grisly maritime anecdote, European colonists sailing to North America in the 16th and 17th centuries were often becalmed in this latitude region and subsequently forced to toss some livestock, typically several horses, overboard in order to conserve the limited amount of drinking water onboard the ship.)

As surface air flows from the subtropical high-pressure regions (horse latitudes) toward the equator, the Coriolis effect deflects it toward the west in both hemispheres to create the northeast and southeast trade winds. Because they blow steadily near the surface, scientists refer to them as prevailing winds. The trade winds from both hemispheres converge at the doldrums.

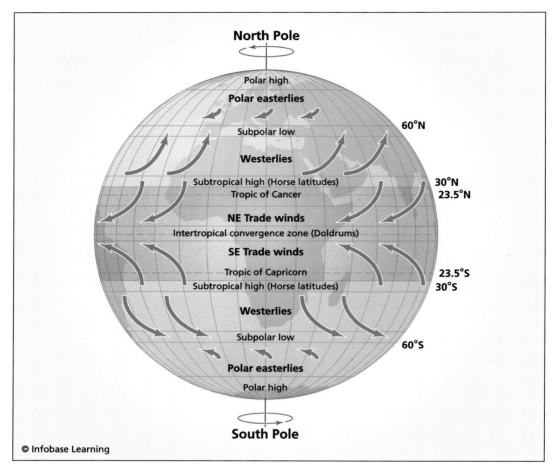

The Coriolis effect influences the circulation pattern of Earth's atmosphere—as depicted here, most of the nonpolar deserts lie within the two trade wind belts. *(USGS)*

Surface winds known as the "westerlies" flow from the horse latitudes toward the poles. (Meteorologists typically name winds by the direction from which they are blowing.) Somewhere between latitudes 50° and 60°N and S, the westerlies encounter the polar easterlies, intermittent winds that blow from the polar high toward the subpolar low (located at about latitude 60°N and S).

Winds can be daily, sporadic, or seasonal. Near Earth's surface, changes in topography and aerodynamic friction influence air flow and the direction of the wind. Winds range from gentle breezes to violent gusts and gales. During extreme weather events such as tornadoes and hurricanes, the wind can attain speeds greater than 186 miles per hour (300 km/hr).

Earth's axial tilt (23.5°) is the reason for the seasons. Because of axis tilt, the Northern and Southern Hemispheres experience opposite seasons. Since Earth's axis points in the same direction (namely at the North Star, called Polaris) all year round, the orientation of the planet relative

PALEOCLIMATOLOGY

Paleoclimatology involves the study of Earth's climate before the widespread use of scientific instruments and the collection of credible temperature, precipitation, and other environmental data. To study the planet's ancient climates, scientists use proxy data. Proxy records of climate have been found preserved in tree rings, extracted as ice cores from glaciers, and discovered buried in laminated sediments found at the bottom of lakes and the ocean. For example, the ages of ice samples found on Earth cover a span that approaches 200,000 years.

One of the most interesting sets of proxy data comes from amber, the fossilized resin of conifer trees. Minute bubbles of ancient air trapped by successive flows of tree resin during the life of the prehistoric tree are preserved in amber. Scientific analyses of the "fossil gases" trapped in these bubbles suggest that Earth's earlier atmosphere during Cretaceous times (some 67 million years ago) contained nearly 35 percent oxygen compared to the present level of 21 percent. The biological consequences of such an elevated oxygen level remain highly speculative.

Astronomers help paleoclimatologists by providing accurate data concerning the very slow wobble of Earth's rotational axis. They refer to this important change in Earth's planetary motion as *precession.* Like a spinning top, Earth precesses (or wobbles) more slowly than it rotates. For Earth's axis, the overall precession cycle takes about 26,000 years. Today, Earth's rotational axis steadfastly points at Polaris, the North Star. Some 13,000 years from now, Earth's axis will point at another star, Vega, which will then serve as the North Star. Other scientific data suggest that the angle of axial tilt varies between 22° and 24° every 41,000 years. Finally, the shape of Earth's orbit around the Sun, called *eccentricity,* appears to change slowly from a nearly perfect circle to an oval shape on a 100,000-year cycle. Such very gradual and very small variations in Sun-Earth geometry alter how much sunlight each hemisphere receives annually, where in each orbit the seasons occur, and how extreme the extent of season changes are.

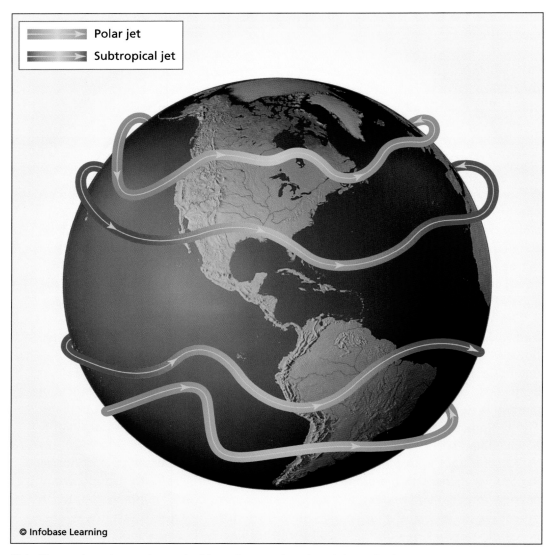

This illustration shows the typical locations and paths of the polar jet stream and the subtropical jet stream for both the Northern and Southern Hemispheres. *(NOAA)*

to the Sun changes as Earth orbits around its parent star. Days are longer and warmer in summer and cooler and shorter in winter. Sunlight strikes the summer hemisphere at a steeper angle than it strikes the winter hemisphere. The steeper the angle, the more concentrated is the incident sunlight.

Jet streams are relatively narrow zones of very strong winds found in the upper troposphere. In jet streams, the winds blow primarily from west to east, but the airflow can often shift to the north or south. Jet streams follow the boundaries between hot and cold air. During winters, hot and cold air boundaries are most pronounced. As a consequence, jet streams are the strongest in winter for both the Northern and Southern Hemispheres.

In discussing jet streams, atmospheric scientists use the terms *higher-latitude polar jet* and *lower-latitude subtropical jet*. They note that jet streams typically vary in height (altitude) from about four miles (6.4 km) to eight miles (12.9 km) and can attain speeds of more than 275 miles per hour (442 km/hr). The actual appearance of jet streams is the result of the complex interactions among many physical variables, such as seasonal weather changes, the presence of warm and cold air, and the location of high and low pressure systems. Jet streams represent a mercurial atmospheric phenomenon. Behaving like "rivers of air," they meander around the globe in both the Northern and Southern Hemispheres. They rise and dip in both altitude and latitude. Sometimes they split and form eddies. At other times, they disappear altogether and then appear somewhere else.

NOAA scientists conclude that jet streams "follow the Sun." As the Sun's elevation increases each day during spring in the Northern Hemisphere, the jet stream shifts north and moves into the atmosphere over Canada by summer. As fall approaches and the Sun's elevation decreases, the jet stream moves south into the United States, where it helps to deliver cooler air to the country.

CLOUDS

Clouds are a distinctive and important feature of Earth's atmosphere. They form when a parcel (bubble) of air gets cooled to its dew point. Atmospheric scientists define the *dew point* as the temperature to which a given parcel of humid air must be cooled at constant pressure in order for the water vapor contained in the air parcel to condense into liquid water, commonly called *dew*. The term *dew point* originates from the fact that during the night, objects near the ground often cool below the dew point temperature and become coated with dew. In a similar fashion, moisture collects around the outside of a glass filled with an icy cold beverage. This occurs because the cold beverage glass chills the surrounding layer of air to the dew point temperature.

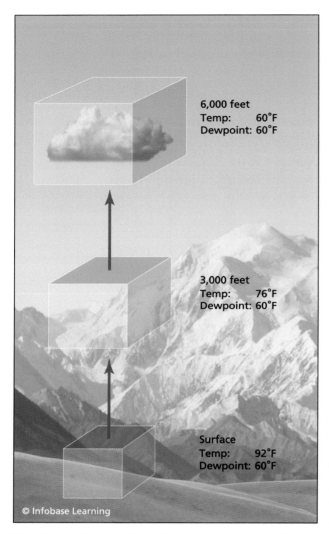

6,000 feet
Temp: 60°F
Dewpoint: 60°F

3,000 feet
Temp: 76°F
Dewpoint: 60°F

Surface
Temp: 92°F
Dewpoint: 60°F

© Infobase Learning

This illustration describes the general process of cloud formation. Clouds form when a bubble, or parcel, of humid air is cooled to its dew point. *(NOAA)*

In thermodynamics, scientists who investigate the vapor-liquid interface use the term *saturation temperature* to designate the temperature at which vaporization takes place at a given pressure, called the saturation pressure. A saturated liquid is one that exists as a liquid at the saturation temperature and pressure. If the temperature of the liquid is lower than the saturation temperature for the existing pressure, then scientists call the liquid a subcooled liquid.

As a parcel (or bubble) of air rises into the atmosphere, it ascends into a region of lower atmospheric pressure. During ascent, the parcel of air expands and cools. Scientists define the lapse rate as the rate of change in temperature experienced while rising through the atmosphere. There are several types of lapse rates used in atmospheric science and meteorology. The one of immediate relevance here is called the dry adiabatic lapse rate. It relates to the temperature changes experienced by a parcel of air as it ascends or descends in the atmosphere. For Earth's atmosphere, the dry adiabatic lapse rate is 5.4°F per 1,000 feet (or 9.8°C per 1,000 m). This means that for each 1,000-foot (304-m) increase in altitude, the air temperature decreases 5.4 °F (3.0°C).

Cold air holds less water vapor than warm air. As an air parcel loaded with moisture rises, some of its water vapor condenses onto tiny particles called condensation nuclei that are suspended in the atmosphere. This is how clouds form. (Aerosols are discussed in chapter 7.) They are the visible aggregate of minute droplets of water, tiny crystals of ice, or combinations of both.

The reverse process takes place when a parcel of air sinks in the atmosphere. As it descends, the parcel encounters increasing pressure, which tends to squeeze (compress) the air and warm it. Since warm air can hold more water vapor than cold air, clouds tend to evaporate when air parcels sink. However, the physics of cloud formation is far more complicated than is briefly discussed in this section.

Atmospheric scientists classify clouds on the basis of their form and height. The three general cloud forms are cirrus, cumulus, and stratus. Cirrus clouds are white, thin, and high and typically form above 20,000 feet (6,100 m). (Cirrus is Latin for "curl of hair" or "filament.") Cirrus clouds generally occur in fair weather and often indicate the direction of air movement at their altitude.

Cumulus clouds resemble large, white, puffy cotton balls. They are a good indicator that thermal uplift (vertical motion) is occurring within the atmosphere. These clouds have a flat base, which indicates the altitude where condensation and cloud formation are taking place. The more humid the air, the lower the altitude of the flat base of a cumulus cloud occurs. The tops of cumulus clouds can reach 60,000 feet (18,300 m).

Stratus clouds cover much of the sky in layers or sheets. (The word *stratus* is Latin for "layer" or "blanket.") The bases of stratus clouds typically occur only a few hundred feet above the ground. The lower layer of these featureless clouds is generally gray, and their presence corresponds to dull weather conditions.

Another basic type of cloud is the nimbus cloud. (Nimbus is the Latin word for "rain" or "rain cloud.") Nimbus clouds typically form between 7,000 feet (2,135 m) and 15,000 feet (4,570 m). They are very dark and bring steady precipitation. When a nimbus cloud thickens and rain begins to fall, its base tends to move closer to the ground.

Scientists use these basic descriptions in various combinations to characterize the 10 cloud forms officially recognized by meteorologists around the world. These are cirrus (Ci), cirrostratus (Cs), cirrocumulus (Cc), altocumulus (Ac), altostratus (As), stratus (St), stratocumulus (Sc), nimbostratus (Ns); cumulus (Cu), and cumulonimbus (Cb). In describing clouds, meteorologists use height categories: high clouds (those above 20,000 feet [6,100 m] altitude), middle clouds (between 6,500 feet [2,000 m] and 20,000 feet [6,100 m]), low clouds (below 6,500 feet [2,000 m]), and clouds that experience vertical development.

Using altitude as the primary figure of merit in categorizing clouds, meteorologists obtain cloud families: high clouds (cirrus, cirrostratus, and cirrocumulus), middle clouds (altocumulus and altostratus), low clouds (stratus, stratocumulus, and nimbostratus), and clouds that develop vertically (cumulus and cumulonimbus). Cumulonimbus clouds are the towering, often anvil-shaped clouds related to severe weather. They generally bring heavy rainfall, thunder, lightning, and hail, and at times spawn tornadoes.

THUNDERSTORMS

This section discusses some of the features of an episode of adverse weather known as a thunderstorm. As the name implies, a thunderstorm is a rain shower during which a person hears thunder. Resembling a sonic boom, thunder is caused by lightning. Meteorologists classify a thunderstorm as severe when it contains one or more of the following: winds gusting in excess of 57.5 miles per hour (92.5 km/h), hail of 0.75 inch (1.91 cm) diameter or greater, or a tornado. According to NOAA severe weather experts, an average thunderstorm has a diameter of 15 miles (24 km) and lasts about 30 minutes. At any given moment, there are approximately 2,000 thunderstorms in progress around the world. This storm frequency results in an estimated 100,000 storms each year on a global basis. Of these storms, only about 10 percent reach severe levels.

The life cycle of a typical thunderstorm has three stages: the developing stage, the mature stage, and the dissipating stage. The developing stage of a thunderstorm is characterized by a cumulus cloud that is being nudged upward by a rising column of air called an updraft. As the updraft continues to develop, a towering cumulus cloud appears. During this stage, some lightning may occur, but there is little or no rain. Although the exact process of lightning formation is still subject to scientific debate, meteorologists suggest that some water vapor in the towering cumulus cloud turns into positively charged ice crystals, while rain droplets generally have negative charges. When there is a sufficient buildup of electric potential, a bolt of lightning becomes the discharge mechanism. As the bolt flashes through the atmosphere, it produces a pressure (sound) wave, which propagates at sonic velocity. The developing stage lasts approximately 10 minutes.

During the mature stage, the updraft continues to feed the storm, but now rain begins to fall. The precipitation is accompanied by the appearance of a column of air that pushes downward. A gusty front, or line of

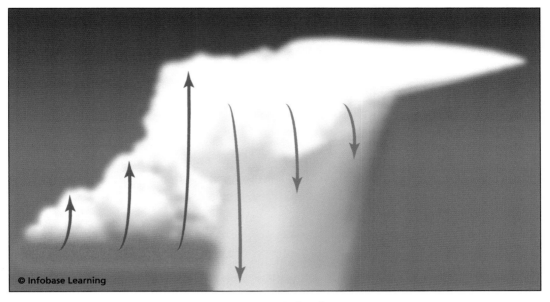

Stages in the development of a thunderstorm *(NOAA)*

gusty winds, forms when the downdraft and rain-cooled air starts spreading out along the ground. Generally, the mature stage involves heavy rain, hail, frequent lightning, strong winds, and the possibility of tornadoes.

The dissipating stage arrives after the storm has produced a large amount of precipitation and the updraft is overcome by the downdraft. The gust front has now moved out a significant distance and starts cutting off the supply of warm, moist air that has been feeding the thunderstorm. Although rainfall intensity decreases, the threat of lightning remains.

Meteorologists categorize thunderstorms as single-cell storms, multicell cluster storms, multicell (squall) line storms, and supercell storms. Each type is described briefly here. A single-cell thunderstorm is uncommon, is not usually severe, and generally lasts less than 30 minutes. Sometimes called a pulse thunderstorm, this type of storm may occasionally produce hail, deliver heavy rainfall for brief periods, generate microbursts (strong downdrafts), or spawn weak tornadoes.

The most common type of thunderstorm is the multicell cluster storm. The storm consists of a group of cells moving along as an atmospheric unit, but with each cell in a different stage of the thunderstorm life cycle. A multicell cluster can persist for several hours and produce flash floods, moderate size hail, and weak tornadoes.

The multicell (squall) line storm consists of a long line of thunderstorms characterized by a continuous, well-developed gust front at the leading edge of the line. Such squall line storms can produce heavy rainfall, golf-ball sized hail, and weak tornadoes.

The supercell storm is rare, highly organized, and an extreme weather event. The updrafts are extremely strong, and air flow can reach speeds of 150 to 175 miles per hour (240 to 280 km/h). The presence of rotation sets the supercell apart from the other kinds of thunderstorms. A supercell's rotating updraft generates giant hail more than two inches (5.1 cm) in diameter, strong downdrafts (microbursts) that exceed 80 miles per hour (129 km/s), and dangerous, very strong, and violent tornadoes.

TORNADOES

A tornado is a violently rotating column of air that extends from a thunderstorm to the ground. With wind speeds of up to 250 miles per hour (400 km/h) or more, the most violent tornadoes can cause a tremendous amount of damage. The path of damage associated with a powerful tornado can exceed one mile (1.6 km) in diameter and 50 miles (80 km) in length.

Although tornadoes can occur in many parts of the world, these destructive severe weather events are found most frequently in the United States and occur primarily east of the Rocky Mountains during the spring and summer months. About 1,000 tornadoes are reported annually by the National Weather Service. On average, tornadoes kill a total of 80 people each year and injure several thousand more.

Most tornadoes are spawned from supercell thunderstorms. As mentioned in the previous section, supercell thunderstorms are characterized by a persistent rotating updraft and form by strong vertical wind shear. Atmospheric physicists describe wind shear as the change in wind speed and/or direction with height. The funnel cloud of a tornado consists of moist air. As the funnel descends from a severe thunderstorm, the water vapor within the funnel condenses into liquid droplets. It is these whirling water droplets that make the funnel cloud visible. Due to the dynamic atmospheric environment, dust and debris on the ground begin to rotate. After the funnel touches the ground and becomes a tornado, its color often changes. Typically the color depends on the type of dirt and debris the tornado passes over.

Tornados can be very brief, lasting only a few seconds, or can persist for an hour. However, most tornados last less than 10 minutes. Severe weather specialists know that the size and shape of a tornado is not a measure of its strength or potential for damage. Gradually, a tornado loses its

The deadly tornado that struck Enterprise, Alabama, on March 1, 2007 *(NOAA)*

intensity. Before dissipation, the condensation funnel gets smaller, and the tornado tilts, exhibiting a contorted, ropelike appearance.

The meteorologist Tetsuya Theodore Fujita (1920–98) developed a scale to estimate tornado wind speeds based on the damage left behind. Today, meteorologists use an improved version of this scale, called the Enhanced Fujita (EF) Scale. On this scale, an EF0 tornado is classified as a weak tornado, with a wind speed of 65 to 85 mph (105 to 137 km/h). The most violent tornado is called an EF5 event. Scientists are uncertain about the upper limit of wind speeds for the strongest tornadoes, so the EF5 tornado is classified as having a wind speed greater than 200 mph (322 km/h).

HURRICANES

Scientists regard hurricanes as the greatest storms on Earth. Few things in nature can compare to the force of these powerful storms. The term *hurricane* derives from Huracan, a god of evil for the ancient Taino, an

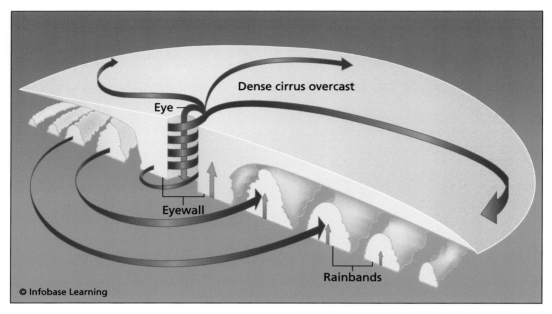

This illustration depicts the main parts and structure of a tropical cyclone (hurricane). *(NOAA)*

aboriginal people who inhabited Central America. In other parts of the globe, people have given these devastating storms other names, such as *typhoon* (in the western Pacific Ocean area) and *cyclone* (in Australia, India, and Pakistan). Regardless of its name, a hurricane is capable of annihilating extended coastal regions with sustained winds of more than 155 mph (250 km/m), intense amounts of rain, and a storm surge.

The scientific name for a hurricane (regardless of its location) is *tropical cyclone*. Generally, a cyclone is a large system of spinning air that rotates around a point of low pressure. Only a tropical cyclone, which has warm air at its center, can become such a powerful storm. Meteorologists use the Saffir-Simpson Hurricane Windscale to classify hurricanes. While useful, this scale addresses only the sustained wind speed of the hurricane and does not take into account the damage potential of other hurricane-related impacts, such as storm surges, rainfall-promoted floods, and tornadoes. The maximum wind speeds for each category of hurricane are: category 1, 74 to 95 mph (119 to 153 km/h); category 2, 96 to 110 mph (154 to 177 km/h); category 3, 111 to 130 mph (178 to 209 km/h); category 4, 131 to 155 mph (210 to 249 km/h); and category 5, greater than 155 mph (>250 km/h). Category 5 hurricanes, though rare, produce incredible levels of destruction.

The figure shows the main parts of a tropical cyclone, including the eye and eyewall. Air spirals in a counterclockwise direction in the Northern Hemisphere (clockwise in the Southern Hemisphere) and out the top in the opposite direction. In the very center of a hurricane, air sinks to form a region called the eye, which is the calmest part of the storm and mostly cloud free.

Typical hurricane-strength tropical cyclones are about 300 miles (483 km) wide, although the dimensions of a particular storm can vary considerably. Some relatively small storms, such as Hurricane Andrew, which hit southern Florida in 1992, are extremely destructive. The hurricane's destructive winds and rains cover a wide swath. Hurricane-force winds can extend outward more than 150 miles (242 km) from the eye of a large storm. The area over which tropical storm-force winds occur is even greater. Tropical-force winds often extend out to almost 300 miles (483 km) from the eye of a large hurricane. The strongest hurricane on record for the Atlantic Basin took place in 2005 and was named Hurricane Wilma.

HARNESSING WIND POWER

Since ancient times, people have harnessed the power of the wind. Early windmills that looked like large paddle wheels first appeared in the Middle East a little more than a millennium ago. Some 200 years later, windmills began to appear in Europe. The primary purpose of these early wind machines, as the

The Sloten Windmill (Molen van Sloten) is situated on the outskirts of Amsterdam, the Netherlands. Dating back to 1847, this polder-draining windmill is still functioning and can pump 15,850 gallons (60,000 L) of water per minute from the polder. *(CIA Factbook)*

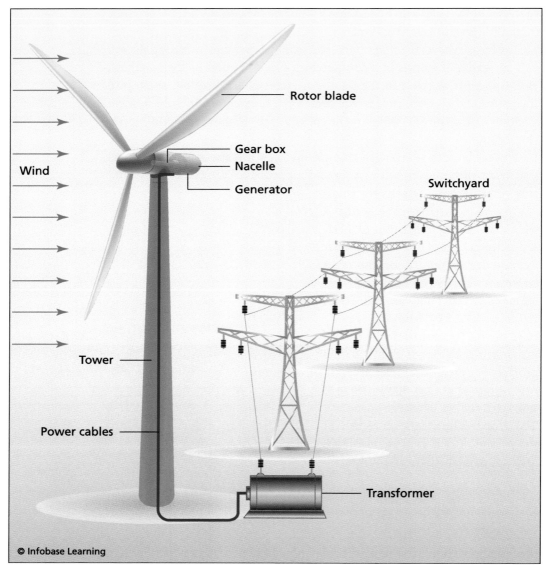

Wind

Rotor blade

Gear box
Nacelle
Generator

Switchyard

Tower

Power cables

Transformer

This diagram shows the main components of a modern wind turbine system designed to harvest the power of the wind and generate electricity. *(TVA)*

name *windmill* implies, was to grind grain. Centuries later, the people of the Netherlands (Holland) improved the design of the basic windmill. They gave it propeller-type blades with which to extract energy from the winds. The blades of the windmill were connected to a drive shaft that operated pumps. In past centuries, the people of Holland used such wind-mills to pump water from polders as part of a major national effort to reclaim additional land from the North Sea.

Today, engineers are designing modern wind machines. These machines, called wind turbines, transform the kinetic energy of naturally flowing air into rotational mechanical energy suitable for the generation of electric power. Most modern wind machines are the horizontal-axis type. Horizontal-axis wind machines are typically very tall (about the height of a 20-story building) and have three blades that span 200 feet (61 m) or more. As the blades of the wind turbine spin, they provide rotary mechanical energy via a gearbox to an electric generator. Although wind energy is a useful renewable energy source, wind turbines cannot produce electricity when the wind is not blowing. Several environmental issues are also being addressed, including noise from the rotating mechanical equipment; the aesthetic (visual) impact of large-scale wind farms that contain 100 or so giant wind turbines; and avian and bat mortality rates, as many of these flying creatures strike rotating blades while attempting to cross a wind farm.

CHAPTER 7

Air Pollution

This chapter discusses the overall problem of air pollution. The World Health Organization (WHO) estimates that air pollution kills more than 8,000 people a day worldwide. There are natural sources of air pollution as well as anthropogenic (human-generated) sources. As discussed in this chapter, volcanic eruptions are not only spectacular displays of a restless planet but are a significant source of natural air pollution. Today, human beings pollute the air with vehicle emissions, gaseous discharges from industrial plants, and the combustion of fossil fuels in large quantities to generate electricity. The often overlooked issue of indoor air pollution is also treated. The chapter concludes with a brief mention of some of the lethal materials intentionally injected into the atmosphere during the testing or use of military weapons. These special anthropogenic pollutants are associated with radiological, chemical, and biological weapons as well as the atmospheric detonation of nuclear weapons.

VOLCANOES

Earth is a restless planet that evolved during the past 4.6 billion years, transitioning from a totally molten world into a habitable planet that possesses a solid outer skin covered by two thin layers of fluids, the oceans and the atmosphere. The planet's surface continues to be influenced by the Sun, gravitational forces, complex interactions with the hydrosphere and atmosphere, and processes emanating from deep within its core.

On very short time scales, people appear to be standing on solid ground, or terra firma. Despite such stable appearances, Earth continuously experiences many geophysical processes that transform and sculpt its surface. These planet-changing processes can be quite dramatic and lead to loss of life as well as involve extensive property damage. Sometimes, these changes happen slowly, such as surface subsidence due to aquifer depletion. Other times, the changes are unpredictable and rapid, such as those brought about by earthquakes, volcanic eruptions, and landslides.

Earth's surface and interior are major components of an interconnected, highly dynamic physical system that scientists call *solid Earth*. Solid-Earth scientists measure both the slow and fast deformations of the planet's surface in order to improve their overall understanding of the most dominant geological processes. While reshaping the surface, volcanic eruptions also pollute the atmosphere, so solid-Earth scientists work with other scientists who are involved in Earth system science or who specialize in the study of air pollution.

A volcano is a vent or opening at the Earth's surface through which magma (molten rock) and associated gases erupt. Scientists also use the term *volcano* to describe the cone that has been built up by effusive (over the top) and explosive eruptions.

Throughout history, volcanoes have represented some of Earth's most powerful and destructive natural forces. On August 24, 79 C.E., the Roman cities of Pompeii and Herculaneum were destroyed when the long-dormant Mount Vesuvius volcano suddenly exploded. Within just a few hours, hot volcanic ash and dust buried the two cities, killing thousands of residents.

The process starts far below the surface as magma accumulates in a hollow portion of the lower strata called a *magma chamber*. Streams of magma then rise from this chamber along a central vent up toward an opening in the surface. Scientists call the raised opening that emits magma the *cone*. When magma erupts onto the surface and flows out of the cone, they refer to the hot, less-viscous flowing material as *lava*. An erupting volcano also emits a stream of rock and ash that scientists call *tephra*.

At temperatures ranging between 1,292°F (700°C) and 2,192° F (1,200°C), lava represents the hottest substance found naturally on Earth's surface. Significantly more viscous than water, lava can flow for a great

The 1947 eruption of the Parícutin volcano in Mexico—a lava-dammed lake appears in the foreground. *(USGS)*

distance before cooling. Once cooled, lava solidifies into igneous rock. Geologists use the term *lava flow* to describe the outpouring of molten rock associated with a nonexplosive volcanic eruption. They refer to very-low-density, highly porous solidified lava as *pumice*. Because of its numerous entrapped gas bubbles and air chambers, pumice floats on water. Scientists often find pumice carried by ocean currents far away from volcanic islands.

Explosive volcanic eruptions are destructive and deadly. During such eruptions, the side or top of the volcano often disappears as clouds of hot tephra blast out into the surroundings. Scientists define a *pyroclastic flow* as a ground hugging avalanche of rock fragments, pumice, hot ash, and escaping volcanic gases that rushes down the side of a volcano at speeds of 60 mph (96 km/h) or more. The temperature within a pyroclastic flow can exceed 932°F (500°C).

Repeated eruptions allow a volcano to grow and form a distinctive shape, based on the molten materials involved in the eruption. Geologists divide terrestrial volcanoes into three general types: stratovolcanoes, cinder volcanoes, and shield volcanoes. Resembling a layer cake with frosting, a stratovolcano forms a symmetrical cone with steep sides as each eruption causes lava and tephra to accumulate in strata, or layers. A cinder volcano is a very small, cone-shaped volcano. When erupting lava blasts into the atmosphere and disintegrates into small pieces, the pieces cool and harden into cinders. The cinders then fall back to Earth and accumulate around the volcano's vent, creating the geologic structure known as a cinder volcano. Finally, a shield volcano forms when successive nonexplosive (effusive) eruptions allow lava to smoothly flow from the cone, spread out across the surface, and produce a volcano with broad, gently sloping sides.

A major explosive eruption injects huge amounts of volcanic gas, aerosol droplets, and ash (pulverized rock and pumice) into the stratosphere. Some volcanic gases, such as carbon dioxide (CO_2), are greenhouse gases that promote global warming, while other gases, such as sulfur dioxide (SO_2), can cause global cooling, ozone (O_3) destruction, and polluted air, known as volcanic smog, or *vog*. Volcanic gas can be harmful to humans, animals, and property. Usually, the hazards from volcanic gases are most severe in the areas that immediately surround an erupting volcano, especially on the flanks downwind of active vents and fumaroles. However, following large eruptions, these hazards can extend for long distances downwind. The April 2010 eruption of the Eyjafjallajökull volcano in

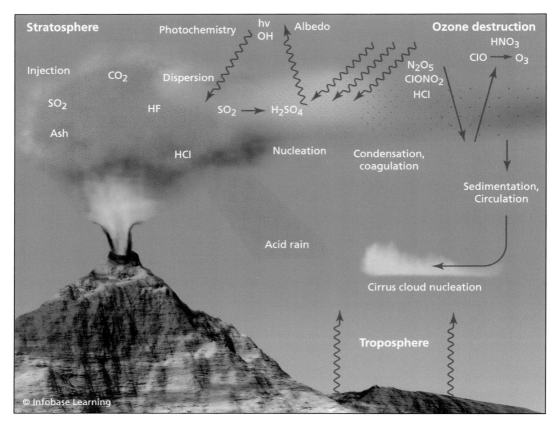

This diagram depicts how an erupting volcano interacts with Earth's atmosphere. *(Adapted from USGS)*

Iceland, for example, disrupted air traffic across northern Europe, stranding thousands of passengers on both sides of the Atlantic. Although the plume of volcanic ash caused massive airline traffic gridlock, an expert group of WHO scientists concluded that the volcanic ash posed no direct threat to public health.

Large, explosive volcanic eruptions inject water vapor (H_2O), carbon dioxide (CO_2), sulfur dioxide (SO_2), hydrogen chloride (HCl), hydrogen fluoride (HF), and ash into the stratosphere to altitudes of 10 to 20 miles (16 to 32 km). The most significant environmental impact from these injections involves the conversion of sulfur dioxide to sulfuric acid (H_2SO_4), which condenses rapidly in the stratosphere to form fine sulfate aerosols. The aerosols (discussed shortly) increase the reflection of incoming solar radiation back into space and therefore cool the lower parts of the atmosphere (troposphere). However, since the aerosols absorb the long-

CAUTION: LUNGS AT WORK!

Many air quality improvement campaigns include signs and posters with the slogan "Caution: Lungs at work!" The phrase is a clever way to remind people of how dependent they are on good quality air for survival. The average person can survive several weeks without food to eat, several days without water to drink, but only minutes without air to breathe. In many regions of the world (especially in the densely populated cities of developing countries), air has become so polluted and foul that people become sick by breathing it.

(continued)

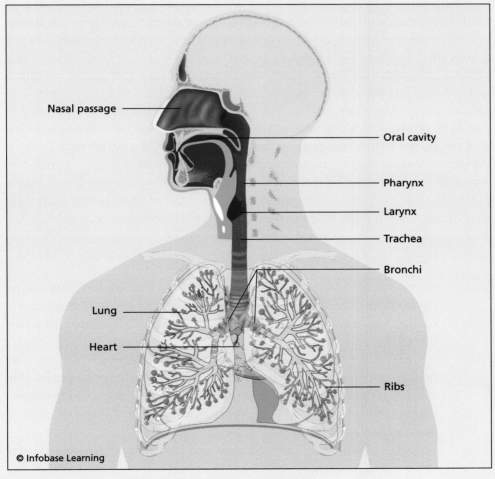

Major parts of the human respiratory system *(DOD)*

(continued)

Respiration is the fundamental physical process by which a living organism exchanges gases with its environment. Breathing is the mechanical portion of respiration, during which human beings and other animals inhale air into the lungs and exhale air out of the lungs. The nasal passage, oral cavity, pharynx, larynx, trachea, and bronchi form the passageway for air to travel from outside the body into the lungs. The lungs serve two primary functions: to acquire oxygen for the body and to remove carbon dioxide from the body. Oxygen is necessary for life; carbon dioxide is a by-product of many life-sustaining chemical reactions within the body.

During breathing, air enters and exits the lungs. It flows into the body through increasingly smaller airways, finally filling tiny sacs called alveoli. Blood pumped by the heart circulates around the alveoli through capillaries (very tiny blood vessels). Where the capillaries and the alveoli meet, incoming oxygen crosses into the bloodstream. Simultaneously, carbon dioxide crosses from the bloodstream into the alveoli to be exhaled. For an adult human being at rest, the overall mechanical act of breathing takes place rhythmically about 12 to 16 times per minute. The brain controls the rate of breathing. A respiratory center located in the brain stem responds to changes in blood chemistry, such as the carbon dioxide burden, as well as stress, the body's motor activities, and changes in temperature.

As part of respiration, a person's lungs are continuously being exposed to particles in the air, including smoke, pollen, dust, and microorganisms. Some of these inhaled substances can cause lung damage or disease if a sufficient quantity is inhaled or if the body is particularly sensitive to them.

wave infrared radiation leaving Earth's surface, their presence warms the stratosphere. The overall balance between aerosol-induced atmospheric cooling and aerosol-induced heating is a subject of contemporary scientific study.

The sulfate aerosols also undergo complex chemical reactions that can alter chlorine and nitrogen chemical species in the stratosphere. This effect, together with increased stratospheric chlorine levels due to chlorofluorocarbon pollution, generates chlorine monoxide (ClO), a compound that destroys ozone (O_3).

As the stratospheric aerosols grow and coagulate, the increased mass causes them to descend into the troposphere. Once in the troposphere, the aerosols serve as condensation nuclei for cirrus clouds. The increase in clouds influences Earth's overall radiation balance.

Most of the hydrogen chloride (HCl) and hydrogen fluoride (HF) molecules are dissolved in water droplets in the volcanic eruption cloud and precipitate quickly to the ground as acid. Most of the ash particles injected into the stratosphere by an explosive eruption descend from the stratosphere within several days to several weeks. Finally, as an integral part of Earth's long-term biogeochemical cycles, volcanic eruptions release large quantities of carbon dioxide, an important greenhouse gas.

Health hazards from the gases associated with large explosive volcanic eruptions range from minor to severe and life threatening. Exposure to acidic gases such as sulfur dioxide, hydrogen sulfide, and hydrogen chloride can damage a person's eyes or the mucus membranes of the respiratory system. Cases of extreme exposure can prove fatal. Prolonged exposure to volcanic smog (vog) produces headache, fatigue, breathing difficulties, and allergic reactions.

Under certain environmental conditions, a very serious hazard can occur from volcanic emissions of carbon dioxide. Since carbon dioxide is 1.5 times heavier than air, it can collect in low and poorly ventilated places. Almost 2,000 people have died of carbon dioxide asphyxiation near volcanoes in the past four decades. Most of the asphyxia victims died in Indonesia or in the Republic of Cameroon. On February 20, 1979, an eruption-released cloud of carbon dioxide overcame 142 villagers in Dieng, Indonesia. Seven additional people died when they entered the area to attempt to rescue the original asphyxia victims. Volcanism in Cameroon created several crater lakes. On August 21, 1986, one of these lakes, Lake Nyos, released a plume of carbon dioxide that killed an estimated 1,746 people. Since the lake is situated high on the flank of an inactive volcano, scientists speculate that a pocket of magma beneath the lake may have leaked a large quantity of carbon dioxide into the water. Volcanic gases can also severely damage vegetation.

OUTDOOR AIR POLLUTION

In addition to natural pollution sources such as volcano eruptions and forest fires, the air people breathe contains emissions from many human-generated sources, such as industry, large-scale agricultural activities, motor vehicles, heating, and commercial sources. The accompanying figure provides a

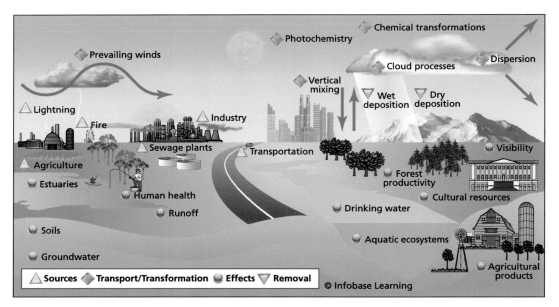

Sources of outdoor air pollution *(Adapted from NPS)*

summary of how these air pollutants travel from their respective sources, wander through the biosphere (mainly due to atmospheric transport), and are removed from the air by chemical transformation or deposition.

Wind and water are great transporters of materials (including pollutants) through Earth's biosphere. Deposition occurs when the transported particles begin falling out of the transporting medium (that is, moving air or water). Scientists use the term *wet deposition* to describe how airborne pollutants return to the surface through rain, snow, fog, or mist. In *dry deposition,* airborne pollutants descend to Earth's surface as particles and gases. As the wind blows caustic particles and acidic gases into contact with trees, buildings, homes, and automobiles, the objects they encounter often experience deterioration. Sometimes, rainstorms combine with dry deposition and lead to aggravated cases of deposited pollution and severe environmental consequences.

WHO and EPA scientists describe air pollutants as airborne particles and gases that occur in concentrations sufficient to endanger the health and well-being of living organisms (especially human beings) or to disrupt the orderly functioning of the environment. Sometimes, the location determines whether a particular particle or gas is treated as a pollutant. If the concentration of helium in a confined workspace gets too high, the

PARADOX OF FIRE

From a scientific perspective, fire is a rapid, persistent chemical reaction that releases heat and light. Fire involves the exothermic (energy-releasing) combination of a combustible substance (fuel) and an oxidant (typically, the gaseous oxygen found in the atmosphere). Chemists define *combustion* as a chemical change, especially an oxidation reaction, accompanied by the release of heat and light.

Fire is both lethal and the key to human survival and progress. Paleoclimatologists speculate that Earth's primitive atmosphere probably resembled the gases released during volcanic eruptions, namely water vapor (H_2O), carbon dioxide (CO_2), sulfur dioxide (SO_2), carbon monoxide (CO), sulfur (S_2), chlorine (Cl_2), nitrogen (N_2), hydrogen (H_2), methane

Distinctive flame resulting from the combustion of a gaseous mixture of hydrogen (77 percent by volume) and methane (23 percent by volume) *(DOE)*

(CH_4), and ammonia (NH_3). There was no free oxygen (O_2), so for about 90 percent of the planet's history, there was no fire on its surface. It was only about 400 million years ago that Earth's atmosphere attained its current level of free oxygen (about 21 percent by volume). Plants on the land (biomass) provided fuel for fire, a sufficient concentration of free oxygen in the air supported combustion, and lightning provided the natural spark. Prehistoric wildfires raged unchecked until they were quenched by rainstorms or limited by natural barriers or self-extinguished due to fuel depletion. Nature remained in charge of fire for the next 398 million years or so.

Then, about 1.5 million years ago, prehistoric humans learned how to make fire and eventually how to control it. Fire gave early humans unprecedented power over their world. A campfire's warmth and light provided protection, and cooking expanded the range and quality of food available for survival.

(continued)

(continued)

Early hunter-gatherers even learned how to use fire to clear the land of scrub vegetation in order to promote the growth of game-attracting vegetation. As the first civilizations emerged, higher-temperature fires, based on the use of charcoal, promoted the production of pottery and metals. Even so, fire often killed people and destroyed early settlements by accident or through acts of warfare.

In just the last 200 years, people have learned how to use fire more efficiently to harvest most of the chemical energy stored in fossil fuels (coal, petroleum, and natural gas). They discovered how to produce very useful quantities of mechanical energy using the steam engine, how to generate large amounts of electricity using giant fossil fuel–fired power plants, and how to travel swiftly through the air using hydrocarbon-fueled jet engines. Scientists understand that the complete combustion of carbon releases thermal energy and also produces carbon dioxide. They also recognize that the incomplete combustion of carbon results in the release of carbon monoxide (CO), soot (primarily unburned particles of carbon), and other atmospheric pollutants.

The use of fire enabled the rise of a global civilization. Many scientists now express concern that the current large-scale use of carbon-fueled fire could so degrade the environment as to hurl modern civilization into an irreversible downward spiral.

normally inert and harmless gas can cause asphyxia, yet large quantities of helium are safely used to fill scientific balloons. (See next chapter.) Scientists regard ozone (O_3) that occurs in the lower portions of the troposphere as a "bad gas," or air pollutant, while ozone that occurs in the stratosphere is called a "good gas" and is *not* treated as a pollutant because at this altitude, the ozone shields the biosphere from the Sun's harmful ultraviolet radiation.

According to EPA and WHO scientists, exposure to air pollution is associated with numerous effects on human health, including pulmonary, vascular, cardiac, and neurological impairments. The health effects vary greatly from person to person. High-risk groups include the elderly, infants, pregnant women, and individuals who suffer from chronic (long-term) heart or lung diseases. Children are often at greater risk to air pollution episodes because they are generally more active outdoors, and their lungs are still developing.

Exposure to air pollution can cause both acute (short-term) and chronic health effects. The acute effects are generally immediate and often reversible once exposure to the air pollutant ceases. Eye irritation, headaches, and nausea are some of the acute health effects associated with air pollution. Chronic effects are usually not immediate, and they tend not to be reversible after exposure ends. Decreased lung capacity and lung cancer are some of the chronic health effects associated with long-term exposure to toxic air pollutants.

Air pollution arises from many different sources. There are human-made stationary sources such as power plants, smelters, and factories, and there are mobile sources, such as automobiles, buses, trucks, planes, and trains. Natural sources include volcanic eruptions, lightning, and

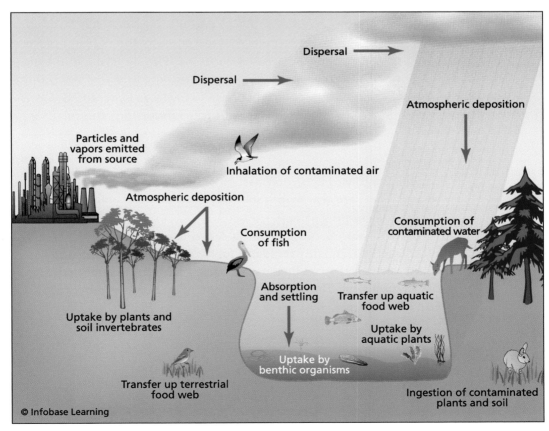

This illustration shows a conceptual model of how air pollutants are transferred within and between ecosystems. *(EPA)*

ACID RAIN

The expression *acid rain* is a broad, general term that refers to a mixture of wet and dry deposition of acidic materials from the atmosphere. By causing freshwater streams and lakes to become more acidic (that is, to have lower pH values), acid rain adversely impacts ecological systems.

Raindrops descending through the air can absorb carbon dioxide (CO_2) and become a dilute solution of carbonic acid (H_2CO_3), a weak acid. Rainwater that has become saturated with carbon dioxide has a natural pH value of 5.6. However, in many parts of the world, rainfall that occurs downwind of industrial sites can become much more acidic and acquire a pH of 3 or less due to interaction with human-caused pollutants. These anthropogenic pollutants include sulfur dioxide (SO_2) from the burning of high–sulfur content coal in electrical generating plants and nitrogen dioxide (NO_2) and nitric oxide (NO) from automobile emissions. The smoke and fumes that accompany the burning of fossil fuels ascend high into the atmosphere and combine with moisture in the air to form mild solutions of sulfuric acid (H_2SO_4) and nitric acid (HNO_3). Precipitation (rain, snow, or fog) brings these pollutants back to Earth as acid rain. Scientists use the term *wet deposition* to describe acidic rain, snow, fog, and mist.

windblown dust. Forest fires can be either human caused or lightning started. There are also significant differences between pollution in industrial and developing nations.

In response to the Clean Air Act, last amended in 1990, the EPA established a set of air quality standards for use in the United States and identified six principal air pollutants considered harmful to public health and the environment. These major air pollutants are carbon monoxide (CO), lead (Pb), nitrogen dioxide (NO_2), particulate matter (PM), sulfur dioxide (SO_2), and ozone (O_3).

Carbon monoxide (CO) is a colorless, toxic gas that forms when carbon in fuel is not burned completely. About 56 percent of all CO in the United States comes from motor vehicle exhaust. The primary sources of lead (Pb) emissions in the United States are non-vehicular equipment, industrial processes, and the combustion of fossil fuels (especially coal).

Nitrogen dioxide (NO_2) is one of a group of highly reactive gases known as oxides of nitrogen, or nitrogen oxides (NO_x). Nitrogen dioxide (NO_2) forms quickly from emissions from automobiles, trucks, buses,

As this acidic water flows over and through the ground, it can adversely affect a wide variety of plants and animals. The magnitude and extent of the environmental consequences depend on several factors, including the acidity (pH value) of the precipitation, the chemistry of the soil, and the ecological systems being stressed. Scientists at the Environmental Protection Agency (EPA) estimate that about 67 percent of all sulfur dioxide (SO_2) and 25 percent of all nitrogen oxides (NO_x) emissions in the United States come from electric power generation that relies on the combustion of fossil fuels such as coal.

In dry deposition, some of the human-produced acidity returns to Earth from the atmosphere as particles and gases. As the wind blows caustic particles and acidic gases into contact with trees, buildings, homes, and automobiles, the objects they encounter often experience deterioration. Sometimes, rainstorms combine acid rain with dry-deposited acid, which leads to aggravated cases of acid deposition and severe environmental consequences. Natural phenomena such as volcanic eruptions and lightning also add acid rain precursor pollutants to the atmosphere. Volcanoes release sulfur dioxides and sulfuric acid, while lightning generates nitrogen oxides and nitric acid.

power plants, and off-road equipment. In addition to contributing to the formation of ground level ozone (so-called bad ozone) and fine particle pollution, NO_2 is linked to a number of adverse effects on the respiratory system.

Particulate matter (PM), or particle pollution, involves a complex mixture of extremely small particles and liquid droplets. Particle pollution components include acids (such as nitrates and sulfates), organic materials, metals, and soil or dust particles.

Sulfur dioxide (SO_2) is one of a group of highly reactive gases known as the oxides of sulfur. The largest human-generated sources of SO_2 emissions are from fossil fuel combustion at power plants (73 percent) and other industrial facilities (about 20 percent). Sulfur dioxide (SO_2) is linked to a number of adverse effects on the respiratory system.

Ozone (O_3) is a paradoxical component of the atmosphere. High in the stratosphere, the oxygen molecule is called "good" ozone, but at ground level, it is "bad" ozone and becomes a major air pollutant. Ground-level ozone is the primary constituent of smog (*smoke* and *fog* combined). Motor

AEROSOL

An aerosol is a very small dust particle or droplet of liquid other than water or ice found in a planetary atmosphere. Aerosols range in size from about 0.001 micrometer (μm) to larger than 100 μm in radius. In Earth's atmosphere, aerosols include smoke, dust, haze, and fumes. They are important as nucleation sites for the condensation of water droplets and ice crystals, as participants in various chemical cycles, and as absorbers and scatterers of solar radiation. Aerosols influence Earth's radiation budget, which in turn influences the climate on the surface of the planet.

Unlike cloud droplets, aerosols can be found in relatively dry air. Natural sources of aerosols include volcanic eruptions and dust storms. Sources of human-generated aerosols include fossil fuel–burning power plants as well as slash-and-burn agricultural practices.

This diagram shows how aerosols in the troposphere can reflect or absorb incoming solar radiation, thereby altering Earth's radiation balance. Aerosols can also have an indirect influence on climate based on the way they interact with surrounding clouds. *(NOAA)*

vehicle exhausts and industrial emissions, gasoline vapors, and chemical solvents—as well as natural sources—emit NO_x and volatile organic compounds (VOCs) that help form ground-level ozone. Sunlight and hot weather can also cause ozone to form in harmful concentrations in the lower atmosphere. Urban areas tend to have high levels of "bad" ozone.

INDOOR AIR POLLUTION

An often overlooked source of air pollution is indoor air pollution. An improperly ventilated home, school, or place of business can literally make a person sick. Indoor air pollution consists of a mixture of contaminants that penetrate a building from outside and those air pollutants generated inside. In developed countries, people spend as much as 90 percent of their time indoors.

As depicted in the figure, there are many sources of indoor air pollution. These include combustion sources (oil, gas, kerosene, coal, wood, tobacco products), building materials, wet or damp carpet, cabinets or furniture made of certain pressed wood products, household cleaning products, central heating and cooling systems, and humidification devices. Air pollution sources found outside the home or office can also influence indoor air quality (IAQ). These include radon and pesticides in addition to all the outdoor air pollutants mention earlier in the chapter.

RADON

WHO scientists report that in many countries, radon exposure is the second-most-important cause of lung cancer after smoking—that is, after exposure to environmental tobacco smoke (ETS). While radon is found both outdoors and indoors, exposure levels remain very low outside a building because of natural dilution in large quantities of air.

Radon (Rn) is a naturally radioactive gas produced by the radioactive decay of radium (Ra), a radioactive element that results from the natural radioactive decay of the thorium (Th) or uranium (U) found in Earth's crust. Radon's most common radioisotope is radon-222, which has a 3.8-day half-life. Radon constitutes a potential health problem for people who occupy improperly designed or poorly ventilated buildings.

Many factors influence indoor air quality in the United States. For example, an improperly adjusted natural gas stove can emit significantly more carbon monoxide than one that is properly adjusted. In developing nations around the world, more than 3 billion people continue to depend on biomass fuels (wood, dung, and agricultural residues) and coal for their energy needs. Unfortunately, carbon monoxide is a significant indoor air pollutant and a silent, deadly killer. Improper ventilation of carbon monoxide from dwellings in such regions causes many fatalities each year, especially among children and women. In developed nations such as the United States, radon and environmental tobacco smoke (ETS) are the two indoor pollutants of greatest concern from a health perspective. Radon is a naturally occurring gas that is odorless, colorless, and radioactive. ETS is the smoke emitted from the burning of a cigarette, pipe, or cigar as

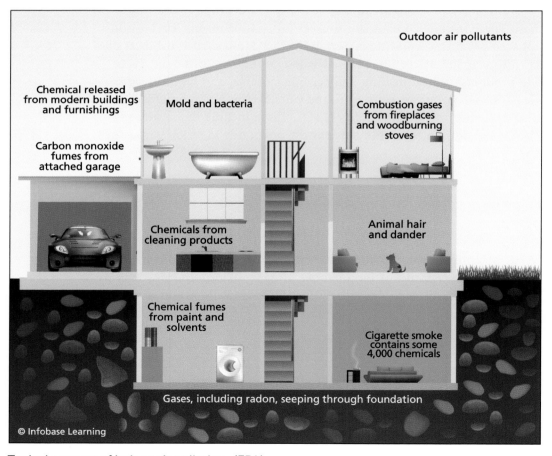

Typical sources of indoor air pollution *(EPA)*

well as the smoke exhaled by a smoker. ETS is a complex mixture of more than 4,000 chemical compounds and contains many known or suspected carcinogens and toxic agents, including particles, carbon monoxide, and formaldehyde.

Microbiological pollution of indoor air comes from hundreds of species of bacteria, fungi, and molds that grow indoors when sufficient moisture is available. Health effects that may arise from chronic exposure to such microbial contaminants include symptoms of respiratory distress, allergies, and asthma.

SPECIAL ANTHROPOGENIC POLLUTANTS

The leaders of rogue nations and terrorist groups threaten global peace with violent acts that may involve the use of nuclear weapons, radiological dispersal devices (so-called dirty bombs), chemical warfare, and biological warfare. Among the harmful consequences of such acts would be the intentional pollution of the atmosphere of a region with radioactive materials, toxic chemicals, or lethal microorganisms. Dispersion in the atmosphere of these harmful, human-produced pollutants would cause not only injury and death but also political and economic disruption and widespread fear. For example, one of the main purposes of a radiological dispersion device would be to spread fear and panic in a highly populated area.

A brief discussion follows regarding the "air pollution" aspects of modern biological warfare. To be used for a maximum credible event, a biological warfare agent would require some special properties. The agent would need to be highly lethal and easily produced in large quantities. Given that the aerosol route is the most likely for a large-scale attack, stability in aerosol and capability to be dispersed (17¼ µm to 5¼ µm particle size) would be necessary. Additional attributes that would make an agent even more dangerous include being communicable from person to person and having no treatment or vaccine.

Scientists at the U.S. Army Medical Research Institute of Infectious Diseases, located at Fort Detrick, Maryland, reviewed the list of potential agents that possess these characteristics and concluded that anthrax and smallpox are the two with greatest potential for mass casualties and civil disruption. Both anthrax and smallpox (variola) are highly lethal: The death rate for anthrax, if untreated before the onset of serious symptoms, exceeds 80 percent; 30 percent of unvaccinated patients infected with variola major could die. There are two major forms of smallpox. Variola major is a severe illness with a high fatality rate; variola minor is much less frequently fatal.

Wearing chemical-biological protective equipment, a U.S. Air Force security team member conducts a search through an industrial facility as part of a combat readiness exercise at Osan Air Base, Republic of Korea, in 1999. *(DOD)*

Both anthrax and smallpox are stable for transmission in aerosol form and capable of large-scale production. Anthrax spores have been known to survive for decades under the right conditions. WHO scientists are concerned that smallpox might be freeze-dried to retain virulence for prolonged periods. Both candidate biological warfare agents have been developed as agents in state-sponsored programs. Iraq under Saddam Hussein had produced anthrax for use in Scud missiles and conducted research on camelpox virus, which is closely related to smallpox. During the cold war, a Soviet defector reported that the former Soviet Union was producing smallpox virus by the ton.

Finally, the use of either anthrax or smallpox could have a devastating psychological effect on the target population and potentially cause widespread panic. This is in part due to the well-documented potential of both biological agents to cause large disease outbreaks. Other potential biological warfare agents include plague and tularemia.

CHAPTER 8

Human Flight

"Man must rise above Earth—to the top of the atmosphere and beyond—for only thus will he fully understand the world in which he lives."

—Socrates, Greek philosopher (469–399 B.C.E.)

This chapter provides a brief summary of how one of humankind's oldest dreams became a reality. From prehistoric times, people have dreamed of flying through the air like the birds. The chapter begins with several ancient legends, explores some of Leonardo da Vinci's visionary notions about flying machines, and recounts the actions taken by pioneering French balloonists in the late 18th century. The science and technology needed to achieve heavier-than-air flight slowly matured in the 19th century. Then, on a windy mid-December day in 1903, all was ready. Two American brothers, Orville (1871–1948) and Wilbur (1867–1912) Wright, ventured from their bicycle shop in Dayton, Ohio, and conducted humankind's first heavier-than-air flight at Kitty Hawk, North Carolina. The chapter also explains how military conflicts and commercial interests drove aeronautical engineering to new heights throughout the 20th century. Technical progress was so rapid that by the start of the 21st century, a fleet of sophisticated aerospace vehicles carried human beings and cargo to an orbiting outpost in space.

Six decades of American military aircraft heritage appear in this unusual photograph of a special four-aircraft flying formation over Nellis Air Force Base, Nevada, on November 12, 2006. Shown (clockwise from top left) are the propeller-driven P-51 Mustang, the F-15 Eagle jet aircraft, the F-16 Fighting Falcon jet aircraft, and the F-22 Raptor stealth aircraft. The F-22 Raptor represents the latest generation of supersonic fighter aircraft. *(USAF)*

THE DREAM OF HUMAN FLIGHT

The study of many ancient cultures reveals that most early peoples reserved the heavens and the gift of flight for their deities. In Greco-Roman mythology, for example, the Olympian gods, such as Zeus (Jupiter) and Hermes (Mercury), dwelled in the heavens and could travel across the sky to visit mortals anywhere on Earth.

One of the most enduring flight legends in Greco-Roman mythology is that of Daedalus and Icarus. Daedalus was the grand architect of King Minos's labyrinth for the Minotaur on the island of Crete. The Minotaur was a ferocious beast with the head of a bull and the body of a man. (In another legend, the Greek hero Theseus unravels thread as he travels into the labyrinth to hunt down the Minotaur and kill the beast.) Following completion of the labyrinth, King Minos imprisoned both Daedalus and his son, Icarus, in a tall tower to protect its secrets. Undaunted, Daedalus, a brilliant engineer, fashioned two pairs of wings out of wax, wood, and leather. Before their aerial escape from the tower, Daedalus cautioned his son not to fly so high that the Sun would melt the wax and cause the wings to disassemble. The pair made good their escape from King Minos's Crete, but while over the sea, Icarus ignored his father's warnings and soared high into the air. Daedalus, who reached Sicily safely, watched his young son, wings collapsed, tumble to his death in the sea below.

There is another interesting legend about flight from imperial China. According to this story, around 1500 C.E., a lesser-known Chinese official named Wan Hu conceived of the idea of flying through the air in a rocket-propelled chair. He ordered the construction of a chair-kite structure to which was attached a collection of 47 fire arrow rockets. (A fire arrow is a long gunpowder rocket invented by the Chinese and used in combat in 1232 C.E. to startle invading Mongol warriors.) Serving as his own test pilot, Wan Hu bravely sat in the chair and ordered his servants to simultaneously light the fuses to all the rockets. All 47 servants, each carrying a small torch, rushed forward at their master's command. Dutifully, they lit the fuses and then dashed back to safety. Suddenly, there was a bright flash and a tremendous roar. The air was filled with billowing clouds of gray smoke. Unfortunately, Wan Hu and his rocket-propelled chair vanished in the explosion.

The Italian Renaissance genius Leonardo da Vinci (1452–1519) carefully watched birds in flight and observed several of the basic principles of aerodynamics. His notebooks are filled with sketches of various flying machines. Science historians generally agree that while the Italian inventor and artist had many interesting designs and concepts, he probably did not attempt to construct prototypes of most of his designs. There are no credible records to support the idea that da Vinci personally attempted to fly using any of his machines. Scientists call flying machines designed with flapping wings to resemble the motion of birds ornithopters. Birds, bats, and insects all fly in this manner.

Da Vinci's numerous sketches of mechanical wings and primitive flying machines influenced other would-be experimenters over the next two centuries. The more daring individuals attempted to imitate how birds fly by attaching various wing configurations to their bodies. Some traveled a few feet before tumbling to Earth; others, who unfortunately jumped off roofs or towers, flapped their mechanical wings hopelessly as they tumbled (much like Icarus) to serious injury or death. However, a few experimenters decided to abandon mechanical wing flapping devices and to pursue the construction of devices that could glide through the air. The most famous of these individuals was the British engineer and aviation pioneer Sir George Cayley (1773–1857).

Some science historians refer to Cayley as the father of aviation or the father of aerodynamics. He was a true pioneer in the history of aeronautics and received credit for the first major breakthrough in heavier-than-air flight. He was the first to identify the four aerodynamic forces of flight (weight, lift, drag, and thrust) and their relationship with Newton's laws of motion. He was also the first to build a successful human-carrying glider. Cayley described many of the concepts and elements of the modern airplane and was the first to understand and explain in engineering terms the concepts of lift and thrust. Before him, researchers thought that the propulsion system should generate both lift and forward motion at the same time, as birds are able to do. Misguided, they constructed their flying machines with flapping wings (ornithopters) to resemble the motions of birds. Cayley realized that the propulsion system should generate thrust, but that the wings should be shaped so as to create lift. Finally, the British aviation pioneer was the first investigator to apply the research methods and tools of science and engineering to solve the problem of heavier-than-air flight.

In his experiments, Cayley first tested his ideas with small models and then gradually progressed to full-scale demonstrations. He also kept meticulous records of his observations.

In 1799, Cayley designed a configuration that was basically in the form of a modern airplane, with a fuselage and wings. A little more than a century before the Wright brothers successfully flew their powered craft, Cayley had established the basic principles and configuration of the modern airplane, complete with fixed wings, fuselage, and a tail unit with elevators and rudder, and had constructed a series of models to demonstrate his innovative concepts.

Experiments that he began to carry out in 1804 allowed him to learn more about aerodynamics and wing structures using a whirling arm device. He observed that birds soared long distances by simply twisting their arched wing surfaces and deduced that fixed-wing machines would fly if the wings were cambered. This was the first scientific testing of airfoils—the part of the aircraft designed to produce lift.

After these experiments, he constructed what some aviation historians consider to be the first real airplane. This glider, which was basically a kite on top of pole, was about five feet (1.5 m) long, with a fixed wing set at an angle of incidence of 6° and a cruciform tail that was attached to the fuselage by universal joints. Movable ballast controlled the center of gravity. After this model successfully flew, Cayley designed a larger model glider with rigid wings.

By 1808, Cayley had constructed a glider with a wing area of almost 300 square feet (28 m²). During the next year (1809), Cayley investigated the improved lifting capacities of cambered wings, the movement of the center of pressure, longitudinal stability, and the concept of streamlining. He also demonstrated the use of inclined, rigid wings to provide lift and roll stability and the use of a rudder steering control. He even came to realize that an area of low pressure is formed above the wing.

In 1783, Daniel Bernoulli had discovered that if the speed of a flowing fluid increases, the pressure decreases. Scientists use Bernoulli's principle to describe how airflow over a wing causes a difference in pressure over the top of the wing versus the air pressure beneath the wing. Specifically, air pressure over the wing is lower than air pressure under the wing. This causes the wing to rise into the pocket of lower-pressure air, causing lift.

By 1809, Cayley had advanced from model gliders to building and successfully flying a glider with a total wing area of approximately 172 square feet (18.5 m²). Soon afterward, Cayley published "On Aerial Navigation" (1809–10), which appeared in Nicholson's *Journal of Natural Philosophy, Chemistry and the Arts*. In this paper, he laid out the basis for the study of aerodynamics. However, this work remained relatively unknown and unacknowledged by other scientists and engineers for some years.

After having built several models (with an interruption to explore the possibility of an aerial carriage in 1843), Cayley concentrated on experiments with full-size gliders. He built his first full-size glider in 1849 and initially carried out trials with ballast. Later that year, the 10-year-old son of one his servants made a short flight in a Cayley glider. With the experiment, the now anonymous young man became the first person in history to fly (or more accurately glide) in a heavier-than-air machine.

In 1853 (50 years before the first powered flight by the Wright brothers), Cayley built a triplane glider (a glider with three horizontal wing structures) that carried his coachman 900 feet (274 m) before crashing. It was the first recorded flight by an adult in a heavier-than-air craft. Throughout his long career, Cayley recognized and searched for solutions to the basic problems of flight. These included the ratio of lift to wing area, the determination of the center of wing pressure, the importance of streamlined shapes, the recognition that a tail assembly was essential to stability and control, the concept of a braced biplane structure for strength, the concept of a wheeled undercarriage, and the need for a lightweight source of power. Cayley correctly predicted that sustained flight would not occur until a lightweight engine was developed to provide adequate thrust and lift, an event that did not take place until the historic powered flight by Orville and Wilbur Wright in 1903.

LIGHTER THAN AIR (LTA) CRAFT

The first public demonstration of a lighter-than-air machine took place on June 4, 1783, in Annonay, France, when Joseph-Michel (1740–1810) and Jacques-Étienne (1745–99) Montgolfier, two brothers who owned a paper mill, sent up an unmanned hot air balloon. They had observed that smoke tended to rise and that paper bags placed over a fire expanded and also rose, pushed upward by the hot air. They concluded that if they could only capture what they thought was a unique gas inside an enclosed light-weight bag, this container or bag would rise from the ground. Étienne Montgolfier carried out the first experiment at Avignon, France, in September 1782, proving their theory to be sound. They had rediscovered Archimedes' theory of buoyancy

Their original test balloon was made of paper and linen and opened at the bottom. When flaming paper was held near the opening, the bag slowly expanded with the hot air and floated upward. The brothers tested balloons ranging in size from 40 cubic feet (1.1 m³) to 650 cubic feet (18.4 m³). The balloons rose from 90 feet (27 m) to 600 feet (183 m) in the air. After concluding that their experiment worked, they finally built a large cloth and paper balloon 30.5 feet (10 m) in diameter and tested it on June 4, 1783, in the marketplace at Annonay. The balloon, from then on called a Montgolfiere, rose about 6,562 feet (2,000 m) into the air.

After their success, the brothers went to Paris and built another larger balloon. On September 19, 1783, in Versailles, the Montgolfiers flew the first passengers in a basket suspended below a hot-air balloon—a sheep, a rooster, and a duck. The flight, which lasted eight minutes, took place in front of King Louis XVI, Queen Marie Antoinette, and the French Court as well as a crowd of about 130,000. The balloon flew nearly 2 miles (3.2 km) before returning the occupants safely to Earth.

The next major milestone occurred on October 15, 1783, when the brothers constructed a hot air balloon that, at the end of a tether, rose 84 feet (25 m) into the air with its first human passengers, Jean-François de Rozier, possibly accompanied by the marquis d'Arlandes. (Historians differ on whether the marquis participated in the first human-crewed balloon flight.) With a capacity of 60,000 cubic feet (17,000 m³), de Rozier stayed aloft for almost four minutes. A short while later, on November 21, 1783, the first confirmed aeronauts, de

Rozier and d'Arlandes, made a free (untethered) ascent in a balloon and flew from the center of Paris to the suburbs, about 5.5 miles (9 km) in some 25 minutes. On January 19, 1784, a huge hot air balloon built by

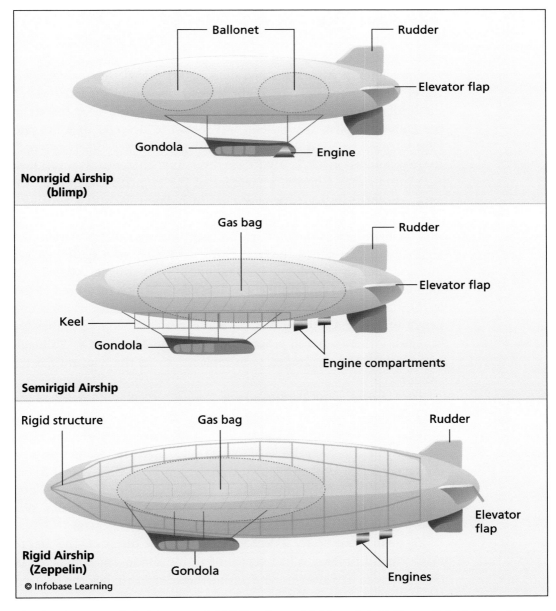

The basic types of airships: nonrigid (blimp), semirigid, and rigid (Zeppelin) *(Adapted from U.S. Centennial of Flight Commission)*

the Montgolfiers carried a total of seven passengers to a height of 3,000 feet (914 m) over the city of Lyons.

At the time, the Montgolfiers believed they had discovered a new gas, which they called "Montgolfier gas," that was lighter than air and caused the inflated balloons to rise. In fact, the gas was simply air that became more buoyant as it was heated. The balloon rose because the contained air was lighter and less dense than the surrounding atmosphere, which pushed against the bottom of the balloon.

Once human passengers had demonstrated that they could safely travel by balloon, balloon flight was firmly established. However, the limitation of using air was soon discovered because, as the air in the balloon cooled, the balloon was forced to descend. If a fire was kept burning to warm the air constantly, sparks were likely to reach the bag and set it afire. Hydrogen overcame this obstacle. Jacques-Alexandre-César Charles, a member of the French Academy of Science, successfully tested a new type of balloon in which hydrogen replaced air in a silk bag that had been treated with an elastic gum, so that hydrogen could not readily pass through it. On August 27, 1783, Charles launched the first balloon inflated with hydrogen gas in Paris. Unlike the Montgolfier balloon, this hydrogen-inflated balloon was closed to contain the gas. The sphere, which measured 13 feet (4 m) in diameter, ascended from the Place des Victories in Paris to a height of nearly 3,000 feet (914 m) and came down some 15 miles (24 km) away, where terrified peasants attacked and destroyed it. Although highly flammable, hydrogen soon replaced air as the buoyant gas for balloons.

The German count Ferdinand von Zeppelin (1838–1917) invented the rigid airship, or dirigible balloon. He was educated at the Ludwigsburg Military Academy and the University of Tübingen. He entered the Prussian army in 1858. Zeppelin went to the United States in 1863 to work as a military observer for the Union army in the American Civil War and later explored the headwaters of the Mississippi River, making his first balloon flight while he was in Minnesota. He served in the Franco-Prussian War of 1870–71 and retired in 1891 with the rank of brigadier general.

Zeppelin spent nearly a decade developing the dirigible. The first of many rigid dirigibles, called zeppelins in his honor, was completed in 1900. Zeppelin flew the world's first untethered rigid airship, the LZ-1, on July 2, 1900, near Lake Constance in Germany, carrying five passengers. The cloth-covered dirigible, which was the prototype of many subsequent models, had an aluminum structure, 17 hydrogen cells, and two 15-horsepower (11.2-kilowatt) Daimler internal combustion engines, each turning two propellers. It was about 420 feet (128 m) long and 38 feet (12 m) in diameter and had a hydrogen gas capacity of 399,000 cubic feet (11,298 m³). During its first flight, the zeppelin flew about 3.7 miles (6 km) in 17 minutes and reached a height of 1,300 feet (390 m). However, it needed more power and better steering. After experiencing technical problems during the flight, the craft was forced it to land in Lake Constance. Following additional tests, the LZ-1 was scrapped three months later.

In 1910, a zeppelin provided the first commercial air service for passengers. By Zeppelin's death in 1917, he had built a zeppelin fleet, some of which were used to bomb London during World War I. However, they were too slow and explosive a target in wartime and too fragile to with-

SCIENTIFIC BALLOONING

NASA's scientific balloon program provides uncrewed, high-altitude platforms for scientific and technological investigations. These investigations include fundamental scientific research that contributes to a better understanding of Earth, the solar system, the Milky Way galaxy, and the universe. Scientific balloons also provide a platform for the demonstration of promising new instruments and spacecraft technologies. When scientists need longer exposure times at high altitudes (typically 120,000 feet [36,600 m]), they use large scientific balloons (about 400 feet [122 m] diameter when full expanded). Scientific balloons are filled with helium and can be launched from almost anywhere. During a typical research

Scientific balloon *(NASA)*

mission, the scientific balloon stays aloft for about 24 hours, and the scientific experiment package is normally tracked and recovered.

stand bad weather. They were found to be vulnerable to antiaircraft fire, and about 40 were shot down over London. After the war, they were used in commercial flights until the explosion and crash of the hydrogen-filled LZ-129 *Hindenburg* at Lakehurst, New Jersey, in 1937.

FIRST POWERED AND CONTROLLED FLIGHT

Influenced by the work of 19th-century aeronautical pioneers such as Sir George Cayley and the German Otto Lilienthal (1848–96), two American brothers, Orville and Wilbur Wright, began an ambitious program of glider construction in their bicycle shop in Dayton, Ohio, at the start of the 20th century. Their primary goal was to demonstrate that heavier-than-air flight was possible.

By June 1903, they had finished designing and building their first powered glider, called *The Flyer*. The aircraft had a wingspan of just over 40

On December 17, 1903, two brothers from Dayton, Ohio, named Wilbur and Orville Wright, were successful in flying an airplane they built. Their powered aircraft flew for 12 seconds above the sand dunes of Kitty Hawk, North Carolina, making them the first to pilot a heavier-than-air machine that took off on its own power, remained under control, and sustained flight. *(NASA)*

feet (12 m), a surface area of 510 feet² (47 m²), and a mass of 625 pounds (283 kg). They constructed as much of *The Flyer* as they could in Dayton and then shipped the remaining parts to Kitty Hawk, North Carolina, for final assembly.

A coin flip decided who would be the aircraft's pilot during this great moment in aviation history. On December 17, 1903, Orville Wright climbed aboard *The Flyer*. Once the restraining wires were released, the craft began to move down a rail into the wind, which was prevalent in Kitty Hawk. The aircraft rose quickly, and humankind's first heavier-than-air flying machine pitched up and down for 12 momentous seconds. *The Flyer* landed with pilot and machine intact save for one damaged skid. Their aircraft had flown a total distance of 120 feet (36.6 m), a distance actually less than the wingspan of modern commercial jet airliners! For the very first time in human history, a powered flying machine had taken off from level ground, traveled through the air using its own propulsion system, and landed under the control of its pilot.

Not satisfied, the Wright brothers made three more successful flights that day. Wilbur was the pilot for the second flight, which traveled a distance of about 175 feet (53 m). Orville made the third flight and landed about 200 feet (61 m) from his starting point. Wilbur was at the controls for the fourth and final flight on that historic day. He kept *The Flyer* on a fairly even course until the aircraft suddenly plunged to the ground. The aircraft had traveled more than 852 feet (260 m) in 59 seconds. Wilbur walked away from the rough landing, but the front rudder of *The Flyer* was badly broken. As the Wright brothers were preparing for still another test flight, a gust of wind flipped *The Flyer* and destroyed the plane. Their little airplane made aviation history that day, but it would never fly again.

JET AIRCRAFT

World War II saw the first use of jet-propulsion aircraft. The first aircraft in the world to fly using turbojet power was the German Heinkel He-178 experimental aircraft. On August 27, 1939, the turbojet airplane was flight tested as a private business venture by the Heinkel Company at Rostock, Germany, just before World War II. Indifference by the senior staff of the German air force caused the project to be abandoned. They expressed no interest in a propellerless aircraft that could achieve speeds of only 375 mph (598 km/h) and operate for only about 10 minutes.

In summer 1944, the German Messerschmitt (Me) 262 (*Schwalbe* ["swallow"]) entered combat service as the world's first operational jet-powered fighter aircraft. With a maximum speed of 559 mph (900 km/h), the Me 262 disrupted Allied bomber formations. However, the aircraft appeared too late in the war to influence the Allied strategic bombing campaigns over Germany or prevent the ultimate defeat of the Nazi regime. Regarded as the most advanced operational German aircraft in World War II, the Me 262 influenced the postwar design of American and Russian (Soviet) military jet aircraft during the first decade of the cold war.

British and American aeronautical engineers and military officials were also investigating the use of jet aircraft. By October 1, 1942, a test pilot flew the first U.S. jet plane, named the Bell P-59A Airacomet, at Muroc, California. Leadership within the U.S. Army Air Forces (AAF) decided not to rush this particular aircraft design into production. Rather, they gathered experience about jet aircraft technology and pursued the development of more advanced flight systems. The British jet fighter aircraft Gloster Meteor made its initial flight in 1943 and became operational for combat in July 1944.

The Lockheed P-80 Shooting Star was the first American aircraft to exceed 500 mph (805 km/h) in level flight, the first American jet airplane manufactured in large quantities, and the first U.S. Air Force jet used in combat. Designed in 1943, the P-80 made its maiden flight on January 8, 1944. U.S. Air Force officials redesignated the aircraft as the F-80 in 1948, when "P" for "pursuit" was changed to "F" for "fighter." Although initially designed as a high-altitude interceptor, the F-80C was flown as a day fighter, fighter-bomber, and photo reconnaissance aircraft during the Korean War. On November 8, 1950, an American F-80C shot down a Russian-built Mig-15 in the world's first all–jet fighter air battle.

After World War II ended and the cold war began, air power developed along a number of peaceful and military lines, each requiring its own technology and machines linked through technology, manufacturers, and transfers of personnel.

The civilian (commercial) airlines benefited enormously from the mass production of aircraft during World War II. Numerous airfields (airports) had also been set up around the world. The London–Los Angeles nonstop reach of big piston-engine airliners was overtaken after 1958 by the arrival of big jets, followed in 1969 by the Boeing 747 jumbo jet. By 1970, ocean liners had been eclipsed by airliners. In the 21st century, larger, more efficient, and higher-capacity jet aircraft speed people and cargo around the planet.

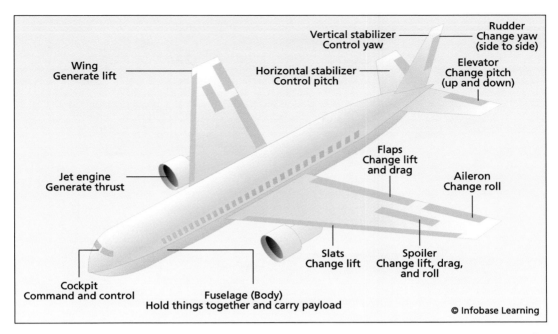

This illustration depicts the major parts and functions of a representative turbine-powered jet airliner. With their ability to efficiently move people and cargo rapidly from one place to another, modern airplanes have transformed the world. *(NASA)*

The figure shows the major components of a modern commercial jet airplane and their functions. Modern airplanes are transportation devices, which aeronautical engineers have designed to efficiently move people and cargo from one place to another. For any aircraft to fly, the vehicle must be able to lift its own weight plus that of the fuel, passengers, and cargo. The wings create most of the lift to hold the plane in the air. The jet engines, located beneath the wings, provide the thrust to push the plane forward through the air. Aerodynamic drag is the atmosphere's resistance to the motion of the jet aircraft. Some planes, typically general aviation aircraft and short-haul commuter aircraft, use propellers for the propulsion system instead of jets.

GAS TURBINES REVOLUTIONIZE AVIATION

To move through the air, an airplane uses some type of propulsion system to generate thrust. The most widely used form of propulsion system for modern aircraft is the gas turbine engine. Jet engines have

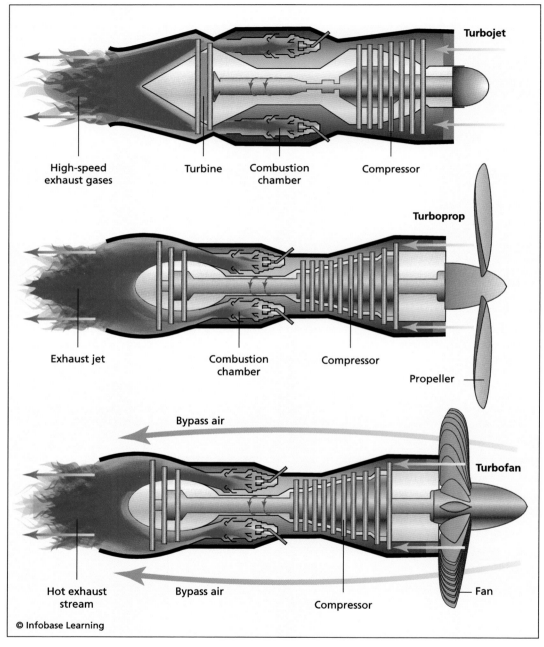

Turbojet

High-speed exhaust gases   Turbine   Combustion chamber   Compressor

Turboprop

Exhaust jet   Combustion chamber   Compressor   Propeller

Bypass air

Turbofan

Hot exhaust stream   Bypass air   Compressor   Fan

© Infobase Learning

The gas turbine engine is the most widely used form of propulsion system for modern aircraft. This figure displays three basic types of gas turbine engines used in aviation. *(NASA)*

continually evolved since their introduction during World War II. The figure shows three different types of gas turbines now used in aviation. Common features of each of these engines include an inlet, a compressor, a combustion section, a turbine, and a nozzle. Aeronautical engineers refer to the compressor, burner, and turbine taken together as the core of the engine, or the gas generator. In the turbojet engine, hot gas passes through the nozzle to produce thrust. In the turbofan and turboprop aircraft engines, the hot gases are used to drive a second turbine that provides rotary power, or torque.

Some high-performance jet engines have an afterburner. An afterburner is a device used for increasing the thrust of a jet engine by burning additional fuel in the uncombined oxygen present in the turbine exhaust gases. Although not fuel efficient, an afterburner provides a military fighter jet the additional thrust necessary to pursue or evade enemy aircraft.

Most modern passenger and military aircraft are powered by gas turbine engines, commonly called jet engines. The first and simplest type of gas turbine is the turbojet. The accompanying figure presents

With its engines roaring, the U.S. Air Force's newest stealth fighter, designated the F-22 Raptor, performs a high-speed aerial maneuver. *(USAF)*

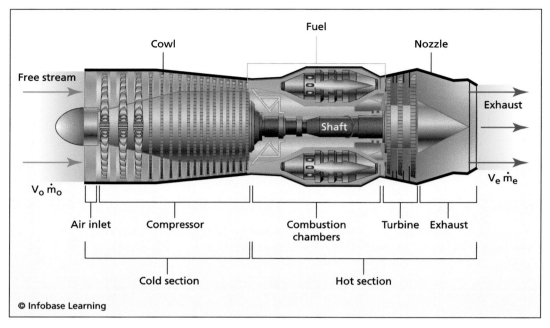

This is a schematic drawing of a basic turbojet engine. *(NASA)*

a simplified schematic drawing of a typical turbojet engine. During operation, large amounts of surrounding air (identified as mass flow rate $\dot{m}_0$) are continuously brought into the engine inlet. At the rear of the inlet, the incoming air enters the compressor. The compressor gathers in air and raises its pressure. In the burner, a small amount of fuel is mixed with the high-pressure air leaving the compressor. A typical modern engine might combine 100 pounds-mass (45 kg) of air per second with 2 lbm (1 kg) of fuel per second. As the hot combustion gases leave the burner, they pass the turbine. The turbine extracts some of the energy in the flowing hot gases. The extracted energy is used to operate the compressor (which is linked to the turbine by a common shaft). The hot combustion gases then pass through a nozzle, which is shaped to expand the exhausting gas to lower pressure and give it a very high exit velocity. The exit velocity (V_e) of the exhausted gas is considerably greater than the air's inlet free stream velocity (V_0). Engineers then determine the net thrust (F) for the turbojet engine from the basic equation $F = \dot{m}_e V_e - \dot{m}_0 V_0$.

ROCKET PLANES TO THE EDGE OF THE ATMOSPHERE

An unofficial motto of flight research in the 1940s and 1950s was "higher and faster." By the late 1950s, the last frontier of that goal was hypersonic flight (Mach 5+) to the edge of space. It required a huge leap in aeronautical technology, life support systems, and flight planning. The North American X-15 rocket plane was built to meet that challenge. It was designed to fly at speeds up to Mach 6 and at altitudes up to 250,000 feet (76,200 m). The aircraft went on to reach a maximum speed of Mach 6.7 and a maximum altitude of 354,200 feet (107,960 m).

The X-15 was a rocket-powered aircraft 50 feet (15.24 m) long with a wingspan of 22 feet (6.71 m). It was a missile-shaped vehicle with an unusual wedge-shaped vertical tail, thin stubby wings, and unique side fairings that extended along the side of the fuselage. The X-15 weighed about 14,000 lbm (6,350 kg) empty and approximately 34,000 lbm (15,420 kg) at launch. The XLR-99 rocket engine, manufactured by Thiokol Chemical Corp., was pilot controlled and was capable of developing 57,000 pounds-force (253,000 N) of thrust. North American Aviation built three X-15 aircraft for the program.

The X-15 rocket plane just after launch from its mother ship in the early 1960s *(NASA)*

X-1 BREAKS THE SOUND BARRIER

The Bell X-1 was a rocket-powered research aircraft patterned on the lines of a 50-caliber machine-gun bullet. It became the first human-crewed vehicle to fly faster than the speed of sound. On October 14, 1947, the Bell X-1 named *Glamorous Glennis* and piloted by Captain Charles "Chuck" Yeager (1923–), was carried aloft by a B-29 bomber mother ship and then released. The pilot ignited the aircraft's rocket engine, climbed, and accelerated, reaching Mach 1.06, or 710 mph (1,142 km/h), as he flew over Edwards Air Force Base in California at an altitude of 8.14 miles (13.1 km). At this altitude, the speed of sound (or Mach 1.0) is 670 mph (1,078 km/h). (The speed of sound in air varies with altitude.) With propellant expended, Yeager glided the aircraft to a landing at Muroc Lake in the Mojave Desert.

The X-15 research aircraft was developed to provide in-flight information and data on aerodynamics, structures, flight controls, and the physiological aspects of high-speed, high-altitude flight. A follow-on program used the aircraft as a platform to carry various scientific experiments beyond Earth's atmosphere on a repeated basis.

For flight in the dense air of the lower portions of the atmosphere, the X-15 used conventional aerodynamic controls, such as rudders on the vertical stabilizers to control yaw and movable horizontal stabilizers to control pitch when moved in synchronization or roll when moved differentially.

For flight in the thin air outside of the appreciable Earth atmosphere, the X-15 used a reaction control system. Hydrogen peroxide thrust rockets located on the nose of the aircraft provided pitch and yaw control. Those on the wings controlled roll.

Because of the large fuel consumption, the X-15 was air-launched from a B-52 aircraft at 45,000 feet (13,715 m) and a speed of about 500 mph (805 km/h). Depending on the mission, the rocket engine provided thrust for the first 80 to 120 seconds of flight. The remainder of the 10- to 11-minute flight was powerless and ended with a 200 mph (322 km/h) glide landing. Generally, one of two types of X-15 flight profiles was used: a high-altitude flight plan that called for the pilot to maintain a steep rate of climb, or a speed profile that called for the pilot to push over and maintain a level altitude.

The X-15 fleet flew a total of 199 flights over a period of nearly 10 years (from June 1959 to October 1968). The aircraft set the world's unofficial speed and altitude records of 4,520 mph (7,273 km/h), or Mach 6.7, and 354,200 feet (107,960 km) in a program to investigate all aspects of piloted hypersonic flight. Information gained from the highly successful X-15 program contributed to the development of the Mercury, Gemini, and Apollo piloted spaceflight programs and also the space shuttle program.

AEROSPACE VEHICLES

An aerospace vehicle is one capable of operating both within Earth's sensible (measurable) atmosphere and in outer space. NASA's space shuttle

A close-up view of the exterior of NASA's space shuttle orbiter *Endeavour* as it travels around the Earth during the STS-123 mission in 2008 *(NASA)*

orbiters are true aerospace vehicles. They leave Earth and its atmosphere under rocket power provided by three liquid-fueled main engines and two solid-propellant boosters attached to an external liquid-propellant tank. After its mission in space is completed, an orbiter vehicle streaks back through the atmosphere and is maneuvered by the crew to land like an airplane. Unlike a modern airplane, however, an orbiter vehicle is without propulsive power during reentry, so the crew lands the vehicle on a runway much like they would a glider.

In 2010, the operational space shuttle fleet consisted of the *Discovery* (OV-103), *Atlantis* (OV-104), and *Endeavour* (OV-105). The *Enterprise* (OV-101) served as a test vehicle but never flew in space. The *Challenger* (OV-99) was lost along with its crew on January 28, 1986, during the first two minutes of the STS 51-L mission; the *Columbia* (OV-102) and its crew were lost on February 1, 2003, at the end of the STS 107 mission while returning to Earth. NASA plans to retire its shuttle fleet in 2011.

CHAPTER 9

Some Interesting Gases

This chapter discusses the characteristics of several important gases. Oxygen and helium represent the third- and second-most-abundant elements in the universe, respectively. Some gases, such as carbon dioxide, are subject to a great deal of technical and political debate. Other gases, such as helium and xenon, have interesting, noncontroversial roles in science. Still others, such as radon, are essentially unnoticed by the general public but pose significant health risks to the general population.

CARBON DIOXIDE

Carbon dioxide (CO_2) is a colorless, odorless, tasteless gas that is about 1.5 times heavier than air. Under normal conditions, this gas is stable, inert, and nontoxic. One of the principal industrial applications of gaseous carbon dioxide is in the production of carbonated beverages. Solid blocks of CO_2 serve as a convenient, portable refrigerant to keep perishable foods cold during transport. Carbon dioxide is also a potent greenhouse gas with a special capability to trap Earth's long-wavelength infrared emissions, thereby raising the temperature of the lower atmosphere and surface.

The combustion of hydrocarbons releases carbon dioxide into the atmosphere. Since the start of the First Industrial Revolution at the end

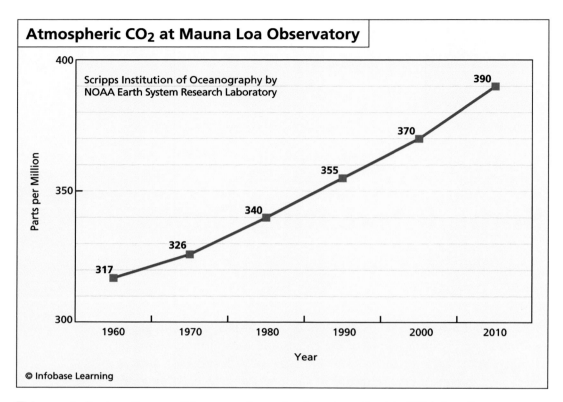

Atmospheric CO_2 at Mauna Loa Observatory

Scripps Institution of Oceanography by
NOAA Earth System Research Laboratory

390

370

355

340

326

317

400

350

300

Parts per Million

1960   1970   1980   1990   2000   2010

Year

© Infobase Learning

This graph displays the monthly mean atmospheric carbon dioxide (CO_2) data (in parts per million) recorded at the Mauna Loa Observatory, Hawaii, from March 1958 to July 2010. *(Scripps Institution of Oceanography/NOAA Earth System Research Laboratory)*

of the 18th century, the atmospheric concentration of certain greenhouse gases has increased. The atmospheric concentration of carbon dioxide, one of the most talked about greenhouse gases, has increased by nearly 30 percent. Other greenhouse gases have also increased in concentrations. These increases have enhanced the heat-trapping ability of Earth's atmosphere. Scientists generally agree that the increase in the concentration of carbon dioxide is due primarily to the combustion of fossil fuels and other human activities.

Carbon is the basis for life on Earth. This essential element moves through the planet's biosphere in a great act of natural recycling called the *global carbon cycle*. Scientists find it helpful to divide the global carbon cycle into two components: the *geological carbon cycle*, which operates over very long time periods (millions of years); and the *physical/biological carbon cycle,* which takes place over much shorter periods (days to millennia).

The carbon that now cycles through Earth's various planetary systems (that is, the biosphere, atmosphere, hydrosphere, and lithosphere) was present in the ancient solar nebula from which the solar system began to form about 5 billion years ago. As Earth emerged as a distinct celestial object about 4.6 billion years ago, the young planet's surface was extensively bombarded by carbon-bearing meteorites. Over time, these

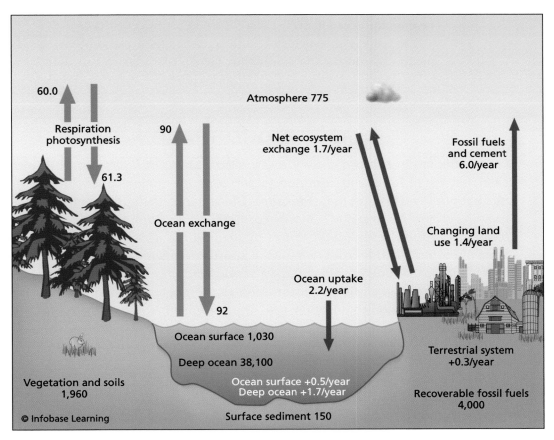

This figure presents a simplified version of the global carbon cycle. The large arrows represent natural paths of carbon exchange, and the small arrows represent anthropogenic (human-caused) contributions to the carbon cycle. Scientists measure the flow of carbon in billions of metric tons (gigatons). *(DOE)*

meteorite impacts steadily increased the planet's carbon content. Since the end of the period of great cosmic collisions, carbonic acid has slowly and steadily combined with calcium and magnesium in Earth's crust to form carbon-containing chemical compounds called carbonates. Carbonic acid (H_2CO_3) is a weak acid that forms when gaseous carbon dioxide (CO_2) dissolves in water (H_2O). Through weathering, the carbonic acid combined with the calcium and magnesium found in Earth's crust to form chemical compounds, such as calcium carbonate (limestone). This process was continual but very gradual. Another natural process, *erosion,* washed the carbonates into the ancient oceans. Once in the ocean, these carbon-bearing compounds precipitated out of the ocean water and formed layers of sediment on the ocean floor.

The process of plate tectonics then pushed these carbon-bearing sediments underneath the continents. Geologists refer to this activity as

CARBON SEQUESTRATION

Energy demand projections by the U.S. Department of Energy in 2010 indicate that fossil fuels will continue to serve as the major source of energy throughout the world for many decades. Since the consumption of fossil fuels remains intimately linked to national security and economic vitality, scientists remain very busy investigating ways of keeping the atmospheric concentrations of carbon dioxide from rising. One approach to controlling carbon emissions from fossil fuels is called *carbon sequestration.*

Scientists define carbon sequestration as the capture and long-term storage of carbon dioxide and other greenhouse gases, such as methane, that would otherwise enter Earth's atmosphere. They suggest the greenhouse gases can either be captured at the point of origin (direct sequestration) or else removed from the atmosphere (indirect sequestration). The captured carbon dioxide can then be stored in underground reservoirs (geological sequestration), injected into deep portions of the oceans (ocean sequestration), or stored in vegetation and soils (terrestrial sequestration).

Geological sequestration involves the storage of captured carbon dioxide in depleted oil and gas reservoirs, in coal seams that can no longer be practically mined, and possibly within underground saline formations. The *subduction.* Once deep within the lithosphere, the limestone and other carbon-bearing sediments experienced increased heat and pressure. The carbonates melted, reacted with other minerals, and released carbon dioxide. Volcanic eruptions then returned the released carbon dioxide to Earth's atmosphere. Scientific evidence suggests that there is a natural balance in the geological carbon cycle between weathering, erosion, subduction, and volcanism. The geological carbon cycle still regulates atmospheric carbon dioxide concentrations—but operates over time periods of hundreds of millions of years.

In the physical/biological carbon cycle, living things play a major role in moving carbon through the biosphere. During the process of *photosynthesis,* green plants absorb carbon dioxide from Earth's atmosphere and use sunlight (energy) to create fuel (glucose and other sugars) for constructing more complex plant structures. Scientists describe

high-pressure injection of carbon dioxide into depleted oil and gas reservoirs may also force any remaining oil or gas toward production wells and facilitate enhanced hydrocarbon recovery operations. Ocean sequestration involves directly injecting carbon dioxide deep into the ocean. Although carbon dioxide is soluble in seawater, and the oceans naturally absorb and release huge amounts of carbon dioxide, this proposed technique is not without controversy. The controversy centers around the impact deep water injection activities might have on the ocean and various marine ecosystems. Terrestrial sequestration involves removal of carbon dioxide from the atmosphere by means of vegetation and soils. Ecosystems that offer significant opportunities for enhanced carbon sequestration include forests, biomass crops, grasslands, and peat lands.

Advanced sequestration concepts, such as converting captured greenhouse gases into rocklike solid materials, are also being investigated. Scientists are examining the feasibility of using minerals such as magnesium carbonate for carbon capture and storage. However, before any carbon sequestration approach is implemented on a large scale, scientists and engineers must demonstrate that the selected approach is technically practical, economic, and environmentally acceptable.

photosynthesis as the natural process through which chlorophyll-bearing green plants use the energy content of sunlight to make carbohydrates from atmospheric carbon dioxide and water. In the basic process, plants transform hydrogen, oxygen, and carbon into the stable organic compound glucose ($C_6H_{12}O_6$) and release water and oxygen back into Earth's environment.

Chemists define *carbohydrates* as any member of a large class of carbon-bearing compounds, including sugars, starches, cellulose, and similar compounds. Carbohydrates serve as the enabling fuel for living things and allow the organisms to grow and reproduce. Plants and animals metabolize (burn) carbohydrates and other nutrient molecules. Through aerobic *respiration* (the opposite of photosynthesis), living organisms release the energy stored in carbohydrates by combining nutrient organic molecules, such as glucose, with oxygen and producing water and carbon dioxide. The released carbon dioxide carries carbon back into the atmosphere. Another natural process, *decomposition* (the digestion of dead or decaying organic matter by bacteria and fungi), also returns carbon fixed by photosynthesis back into the atmosphere. Carbon circulates through the biosphere because of the linkage between photosynthesis, respiration, and decomposition.

Photosynthesis and respiration play important roles in moving carbon through the biosphere—on time scales far shorter than those involved in the geological carbon cycle. The biological processes for various living organisms are elegantly complex; only the very basic carbon exchange activities are mentioned here. Scientists currently estimate that the yearly quantity of carbon fixed by photosynthesis and released back to the atmosphere by respiration is about 1,000 times greater than the amount of carbon that moves through the geological carbon cycle on an annual basis.

Millions of years ago, some of the carbon involved in biological processes was not released back into the atmosphere as carbon dioxide. Instead, buried deposits of dead plants on land and certain marine life forms in the oceans became compressed over time by layers of sediment and eventually formed fossil fuels such as coal, oil, and natural gas. The carbon locked in these fossil fuels remained trapped within Earth's crust for millions of years—until humans mined the various fuels and began burning them. Since the start of the Industrial Revolution in the 17th century, the carbon dioxide content of Earth's atmosphere has increased from about 280 parts per million (ppm) to a current-day (2010) value of about 385 ppm. Human activities, especially the consumption of enormous quantities of fossil fuels and large-scale deforestation, now exert a measurable influence on the planet's global carbon cycle. Many scientists are alarmed at the increased quantity of carbon dioxide in the atmosphere and warn about the dire consequences of global warming due to greenhouse gas build-up. Climate models indicate that increased greenhouse gas concentrations, including methane (CH_4) and carbon dioxide, have been the primary driver of Earth's increasing surface temperature. Improved ways of tracing the amount and pathways of carbon as this element travels throughout Earth's biosphere is an important aspect of Earth system science. Such scientific activities should provide data that lead to a more responsible level of stewardship of humans' home planet in this century.

OXYGEN

Oxygen had been produced by several chemists before its discovery in 1774, but they failed to recognize it as a distinct element. Joseph Priestley and Carl Wilhelm Scheele both independently discovered oxygen, but Priestley is usually given credit for the discovery. They were both able to produce oxygen by heating mercuric oxide (HgO). Priestley called the gas produced in his experiments "dephlogisticated air," and Scheele called his "fire air." (The name *oxygen* was created by Antoine Lavoisier, who incorrectly believed that oxygen was necessary to form all acids.)

Oxygen has an atomic number of 8, an atomic weight of 15.9994, a melting point of −361.82°F (−218.79°C [54.36 K]), and a boiling point of −297.31°F (−182.95°C [90.20 K]).

Oxygen is the third-most-abundant element in the universe and makes up nearly 21 percent of the Earth's atmosphere. Oxygen accounts for nearly half of the mass of Earth's crust, two-thirds of the mass of the human body, and nine-tenths of the mass of water. Large amounts of oxygen can be extracted from liquefied air through a process known as fractional distillation. Oxygen can also be produced through the electrolysis of water or by heating potassium chlorate ($KClO_3$).

Oxygen is a highly reactive element and is capable of combining with most other elements. It is required by most living organisms and for most forms of combustion. Impurities in molten pig iron are burned away with streams of high-pressure oxygen to produce steel. Oxygen can also be combined with acetylene (C_2H_2) to produce an extremely hot flame used for welding. Liquid oxygen, when combined with liquid

GOOD OZONE AND BAD OZONE

Ozone (O_3) is a gas composed of three oxygen atoms. Ozone has the same chemical structure whether it is present miles above the Earth in the stratosphere or near street level in a large city. Ozone is deemed "good" or "bad" by its location. Bad ozone is ground-level ozone and is the primary constituent of smog.

Good ozone occurs naturally in the stratosphere (about 10 to 30 miles [16 to 48 km] above Earth's surface) and forms a protective layer for living things below in the troposphere against the Sun's harmful ultraviolet (UV) radiation. Scientists have linked ultraviolet radiation from the Sun to various types of skin cancer and cataracts as well as damage to some crops, certain materials, and several forms of marine life.

In the mid-1970s, environmental scientists discovered that some human-produced gases could cause stratospheric ozone depletion. Gases containing chlorine and bromine accumulate in the lower atmosphere and are eventually transported into the stratosphere. Once in the stratosphere, these gases convert to even more reactive gases that participate in ozone destroying reactions. Ozone depletion allows additional UV radiation to pass through the atmosphere and reach Earth's surface. This increase in UV radiation exposure causes an increase in adverse health and environmental effects.

Scientists have associated several human-released substances with the depletion of ozone in the stratosphere. These substances include chlorofluorocarbons (CFCs), carbon tetrachloride, methyl bromide, and methyl chloroform.

(opposite page) Role of chlorofluorocarbons (CFCs) in stratospheric ozone (O_3) depletion *(EPA)*

hydrogen, makes an excellent rocket fuel. Ozone (O_3) forms a thin, protective layer around the Earth that shields the surface from the Sun's ultraviolet radiation. Oxygen is also a component of hundreds of thousands of organic compounds.

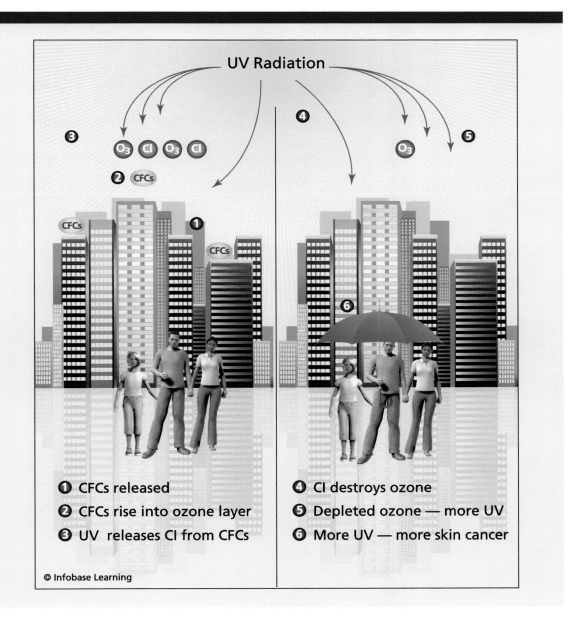

UV Radiation

❶ CFCs released
❷ CFCs rise into ozone layer
❸ UV releases Cl from CFCs
❹ Cl destroys ozone
❺ Depleted ozone — more UV
❻ More UV — more skin cancer

© Infobase Learning

NITROGEN

Nitrogen was discovered by the Scottish physician Daniel Rutherford in 1772. It is the fifth-most-abundant element in the universe and makes up about 78 percent of Earth's atmosphere, which contains an estimated 4,000 trillion tons of the relatively inert gas. Nitrogen is obtained from liquefied air through a process known as fractional distillation.

Nitrogen has an atomic number of 7, an atomic weight of 14.0067, a melting point of −346.00°F (−210.00°C [63.15 K]), and a boiling point of −320.44°F (−195.79°C [77.36 K]).

The largest use of nitrogen is for the production of ammonia (NH_3). The Haber process combines large quantities of nitrogen with hydrogen to produce ammonia. Large amounts of ammonia are then used to create

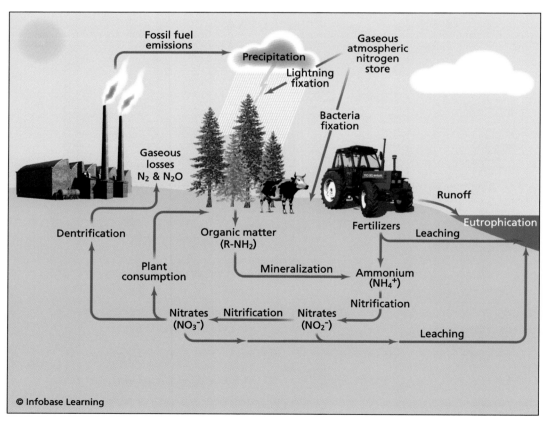

The nitrogen cycle

fertilizers, explosives, and, through a method known as the Ostwald process, nitric acid (HNO_3).

Nitrogen gas is largely inert and is used as a protective shield in the semiconductor industry and during certain types of welding and soldering operations. Oil companies use high-pressure nitrogen to help force crude oil to the surface. Liquid nitrogen is an inexpensive cryogenic liquid used for refrigeration, for the preservation of biological samples, and for low-temperature scientific experimentation

Nitrogen is a component of many organic chemicals. Living things use nitrogen to make a number of complex organic molecules, such as nucleic acids, amino acids, and proteins. The nitrogen cycle is one of the biosphere's most important cycles. Tiny microorganisms help complete this cycle by using nitrate (NO_3^-) as their oxygen source in the decomposition of organic matter and the release of gaseous nitrogen (N_2) back into the atmosphere.

AMMONIA

Ammonia (NH_3) is a colorless alkaline gas characterized by a pungent odor and acrid taste that is highly soluble in water. Ammonia gas may be released from decomposing manure and urine as well as from artificial fertilizers. The toxic chemical has a melting point of –108°F (–77.7°C [195K]) and a boiling point of –28°F (–33.3°C [240 K]). Because of its favorable vaporization properties, engineers often use ammonia as the working fluid in industrial-scale refrigeration systems. At standard temperature and pressure, ammonia is less dense (lighter) than air.

Known to Muslim and European alchemists, Joseph Priestley was the first modern scientist to isolate ammonia. He accomplished the feat in 1774 and named the newly discovered gas alkaline air. During World War I, the German chemist Fritz Haber (1868–1934) invented a process for the manufacture of ammonia from atmospheric nitrogen (N_2) and hydrogen (H_2). He received the 1918 Nobel Prize in chemistry for this work. Haber's process for turning atmospheric nitrogen into ammonia-based agricultural fertilizers helps a large portion of the world's population grow enough food for survival. Haber, a loyal German patriot, was also the chemist who personally supervised the use of chemical weapons against Allied soldiers during World War I. On April 22, 1915, for example, the German army used chlorine gas released from cylinders to kill

Controlled release of ammonia (NH₃) at the Ketchikan Landfill, Alaska *(ADEC-Y. Ha)*

5,000 Allied troops and permanently injure an additional 10,000 during the Second Battle of Ypres in Belgium.

FLUORINE

Fluorine is the most reactive of all elements, and no chemical substance is capable of freeing fluorine from any of its compounds. For this reason, fluorine does not occur freely in nature and was extremely difficult for scientists to isolate. The first recorded use of a fluorine compound dates to around 1670. Information regarding this use of fluorine involves a set of instructions for etching glass that called for Bohemian emerald (CaF_2). The French chemist Ferdinand-Frédéric-Henri Moissan (1852–1907) was the first to successfully isolate fluorine in 1886. He accomplished this through the electrolysis of potassium fluoride (KF) and hydrofluoric acid (HF). He also completely isolated fluorine gas from hydrogen gas, and he built his electrolysis device completely from platinum. He received the Nobel Prize for chemistry in 1906 for his work. Today, fluorine is still produced through the electrolysis of potassium fluoride and hydrofluoric acid as well as through the electrolysis of molten potassium acid fluoride (KHF_2).

Fluorine has an atomic number of 9, an atomic weight of 18.9984, a melting point of −363.32°F (−219.62°C [53.53 K]), and a boiling point of −306.62°F (−188.12°C [85.03 K]).

Fluorine is a corrosive, pale yellow gas. Public health officials often add fluorine to city water supplies in the proportion of about one part per million to help prevent tooth decay. Sodium fluoride (NaF), stannous (II) fluoride (SnF_2), and sodium monofluorophosphate (Na_2PO_3F) are fluorine compounds added to toothpaste to help prevent tooth decay. Hydrofluoric acid (HF) is used to etch glass, including most of the glass used in light bulbs. Uranium hexafluoride (UF_6) is used in the gaseous diffusion process to separate isotopes of uranium. Crystals of calcium fluoride (CaF_2), also known as fluorite and fluorspar, are used to make lenses to focus infrared light. Fluorine joins with carbon to form a class of compounds known as fluorocarbons. Some of these compounds, such as dichlorodifluoromethane (CF_2Cl_2), were widely used in air conditioning and refrigeration systems and in aerosol spray cans but have been phased out due to the damage they were causing to Earth's ozone layer.

CHLORINE

Since it combines directly with nearly every element, chlorine is never found free in nature. Chlorine was first produced in 1774 by the Swedish chemist Carl Wilhelm Scheele when he combined the mineral pyrolusite (MnO_2) with hydrochloric acid (HCl). Although Scheele thought the gas produced in the experiment contained oxygen, Sir Humphry Davy proved in 1810 that it was actually a distinct element. Today, most chlorine is produced through the electrolysis of aqueous sodium chloride (NaCl).

Chlorine has an atomic number of 17, an atomic weight of 35.453, a melting point of −150.7°F (−101.5°C [171.65 K]), and a boiling point of −29.27°F (−34.04°C [239.11 K]).

Chlorine is commonly used as an antiseptic and is used to make drinking water safe and to treat swimming pools. Large amounts of chlorine are used in many industrial processes, such as in the production of paper products, plastics, dyes, textiles, medicines, antiseptics, insecticides, solvents, and paints.

Two of the most familiar chlorine compounds are sodium chloride (NaCl) and hydrogen chloride (HCl). Sodium chloride, commonly known as table salt, is used to season food and in some industrial processes. Hydrogen chloride, when mixed with water (H_2O), forms hydrochloric acid, a strong and commercially important acid. Other chlorine compounds include chloroform ($CHCl_3$), carbon tetrachloride (CCl_4), potassium chloride (KCl), lithium chloride (LiCl), magnesium chloride ($MgCl_2$), and chlorine dioxide (ClO_2).

Chlorine is a very dangerous material. Liquid chlorine burns the skin, and gaseous chlorine irritates the mucus membranes. Concentrations of the greenish-yellow gas as low as 3.5 parts per million can be detected by smell, while concentrations of 1,000 parts per million can be fatal after just a few deep breaths.

HELIUM

Helium (He), the second-most-abundant element in the universe, was discovered on the Sun before it was found on Earth. The French astronomer Pierre-Jules-César Janssen (1824–1907) noticed a yellow line in the Sun's spectrum while studying a total solar eclipse in 1868. The British physicist Sir Norman Lockyer (1836–1920) realized that this line, with a wavelength of 587.49 nanometers, could not be produced by any element then known. It was hypothesized that a new element on the Sun was responsible for this mysterious yellow emission. Lockyer named this unknown element helium after Helios, the sun-god in Greek mythology.

The hunt to find helium on Earth ended in 1895. The Scottish chemist Sir William Ramsay (1852–1916) conducted an experiment with a mineral that contains uranium called clevite. He exposed the clevite to mineral acids and collected the gases that were produced. He then sent a sample of these gases to two scientists, Lockyer and Sir William Crookes, who were able to identify the helium within it. Two Swedish chemists, Nils Langlet and Per Theodor Cleve, independently found helium in clevite at about the same time as Ramsay.

Helium is an inert gas and does not easily combine with other elements. There are no known compounds that contain helium, although attempts are being made to produce helium diflouride (HeF_2).

Helium has an atomic number of 2 an atomic weight of 4.002602, a melting point of −458.0°F (−272.2°C [0.95K]), and a boiling point of −452.07°F (−268.93°C [4.22 K]).

Helium makes up about 0.0005 percent of Earth's atmosphere. This trace amount is not gravitationally bound to the Earth and is constantly lost to space. The helium in Earth's atmosphere is replaced by the decay of radioactive elements in Earth's crust. Alpha decay is a type of radioactive decay that produces alpha particles. An alpha particle can become a helium atom once it captures two electrons from its surroundings. This newly formed helium eventually works its way to the atmosphere through cracks in the crust.

Helium is commercially recovered from natural gas deposits located primarily in Texas, Oklahoma, and Kansas. Helium gas is used to inflate blimps, scientific balloons, and party balloons. It is used as an inert shield for arc welding, to pressurize the fuel tanks of liquid-fueled rockets, and in supersonic wind tunnels. Helium is combined with oxygen to create a nitrogen-free atmosphere for deep sea divers so they will not suffer from a condition known as nitrogen narcosis. Liquid helium is an important cryogenic material and is used to study superconductivity and to create superconductive magnets.

NEON

Neon (Ne) was discovered in 1898 by the Scottish chemist Sir William Ramsay and the British chemist Morris William Travers (1872–1961) shortly after their discovery of the element krypton. Like krypton, neon was discovered through the study of liquefied air. Although neon is the fourth-most-abundant element in the universe, only 0.0018 percent of Earth's atmosphere is neon.

Neon has an atomic number 10, an atomic weight of 20.1797, a melting point of −415.46°F (−248.59°C [24.56K]), and a boiling point of −410.94°F (−246.08°C [27.07 K]). The name *neon* comes from the ancient Greek word for "new."

Bright neon signs welcome evening visitors to a performance at Radio City Music Hall in New York City. *(CIA Factbook)*

The largest use for neon gas is in advertising signs. Neon is also used to make high-voltage indicators and is combined with helium to make helium-neon lasers. Liquid neon is used as a cryogenic refrigerant. Neon is highly inert and forms no known compounds, although there is some evidence that it could form a compound with fluorine.

ARGON

Argon (Ar) was discovered in 1894 by the Scottish chemist Sir William Ramsay and the British physicist Lord John William Rayleigh (1842–1919). The name *argon* derives from the ancient Greek word for "inactive" or "lazy." Argon makes up 0.93 percent of Earth's atmosphere, making it the third-most-abundant gas (after nitrogen and oxygen). Argon is obtained from air as a by-product of the production of oxygen and nitrogen.

Argon has an atomic number of 18, an atomic weight of 39.948; a melting point of −308.83°F (−189.35°C [83.80 K]), and a boiling point of −302.53°F (−185.85°C [87.3 K]).

Argon is frequently used when an inert atmosphere is needed. It is used to fill incandescent and fluorescent light bulbs to prevent oxygen from corroding the hot filament. Argon is also used to form inert atmospheres for arc welding, growing semiconductor crystals, and processes that require shielding from other atmospheric gases.

Once thought to be completely inert, argon is known to form at least one compound. The synthesis of argon fluorohydride (HArF) was reported in August 2000. Stable only at very low temperatures, argon fluorohydride begins to decompose once it warms above −411°F (−246°C). Because of this limitation, argon fluorohydride does not appear to have any uses outside of basic scientific research.

KRYPTON

Krypton (Kr) was discovered on May 30, 1898, by the Scottish chemist Sir William Ramsay and the British chemist Morris William Travers while they were investigating liquefied air. The name of this inert gas derives from the ancient Greek word meaning "hidden." Small amounts of liquid krypton remained behind after the more volatile components of liquid air had boiled away. Earth's atmosphere is about 0.0001 percent krypton.

Krypton has an atomic number of 36, an atomic weight of 83.798, a melting point of −251.25°F (−157.36°C [115.79 K]), and a boiling point of −243.80°F (−153.22°C [119.93 K]).

The high cost of obtaining krypton from air has limited its practical applications. Krypton is employed in some types of photographic flashes used in high-speed photography. Some fluorescent light bulbs are filled with a mixture of krypton and argon gases. Krypton gas is also combined with other gases to make luminous signs that glow with a greenish-yellow light. In 1960, the international scientific community defined the length of the meter in terms of the orange-red spectral line of krypton-86, an isotope of krypton.

Once thought to be completely inert, krypton is known to form a few compounds. Krypton difluoride (KrF_2) is the easiest krypton compound to make, and gram amounts of it have been produced.

XENON

Xenon was discovered by Sir William Ramsay, a Scottish chemist, and Morris M. Travers, an English chemist, on July 12, 1898, shortly after their

A xenon-fueled ion engine operating in a vacuum chamber at NASA's Jet Propulsion Laboratory *(NASA/JPL)*

discovery of the elements krypton and neon. Like krypton and neon, xenon was discovered through the study of liquefied air. Earth's atmosphere is about 0.0000087 percent xenon.

Xenon has an atomic number of 54, an atomic weight of 131.293, a melting point of −169.22°F (−111.79°C [161.36 K]), and a boiling point of −162.62°F (−108.12°C [165.03 K]).

Xenon produces a brilliant white flash of light when it is excited electrically and is widely used in strobe lights. The light emitted from xenon lamps is also used to kill bacteria and to power ruby lasers. Due to its high atomic weight, xenon ions were used as a fuel in an experimental ion engine aboard NASA's space probe *Deep Space 1*.

Once thought to be completely inert, xenon does form compounds, usually with fluorine, oxygen, and platinum. $XePtF_6$, XeF_2, XeF_4, XeF_6, and XeO_4 are some of the xenon compounds that have been formed.

RADON

Radon was discovered by Friedrich Ernst Dorn, a German chemist, in 1900 while studying radium's decay chain. Originally named niton (after *nitens,* the Latin word for "shining"), it has been known as radon since 1923. Today, radon is still primarily obtained through the decay of radium.

Radon has an atomic number of 86, an atomic weight of 222, a melting point of −96°F (−71°C [202 K]), and a boiling point of −79.1°F (−61.7°C [211.45 K]). It is also naturally radioactive.

At normal room temperatures, radon is a colorless, odorless, radioactive gas. The most common isotopes of radon decay through alpha decay. Alpha decay usually is not considered a great external radiological hazard, since the alpha particles produced by the decay are easily stopped. However, since radon is a gas, it is easily inhaled, and living tissue is thereby directly exposed to the radiation. Although it has a relatively short half-life, radon decays into longer-lived, solid, radioactive elements that can collect on dust particles and be inhaled. For such reasons, health officials express concern about the amount of radon present within homes and workplaces. Radon seeps into houses as a result of the decay of radium, thorium, or uranium ores underground and varies greatly from location to location. On average, Earth's atmosphere is 0.0000000000000000001 percent radon.

When cooled to its solid state, radon glows yellow. The glow becomes orange-red as the temperature is lowered. Radon's longest-lived radioisotope, radon-222, has a half-life of about 3.8 days. It decays into polonium-218 through alpha decay. Small amounts of radon are sometimes used by hospitals to treat some forms of cancer. Radon fluoride (RnF) is the only confirmed compound of radon.

CHAPTER 10

Gases for Energy

This chapter discusses how gases form an integral part of today's global energy structure and how humankind's energy future on the planet may hinge on two interesting sources: ice that burns and the most abundant element in the universe. Methane hydrates are the ice that burns, while hydrogen not only was the first element formed as a result of the big bang some 13.7 billion years ago and the most abundant one, it also powers the stars in the visible universe and represents an important energy option this century and beyond.

NATURAL GAS

Natural gas is mostly a mixture of methane (CH_4), ethane (C_2H_6), and propane (C_3H_8), with methane making up about 73 to 95 percent of the total. Often found during drilling for petroleum, natural gas was once treated as a nuisance and burned off. Today, natural gas is mostly injected back into the ground for later recovery and use. When injected under pressure back into an oil formation, natural gas can also stimulate petroleum recovery and production.

The main ingredient in natural gas, methane is a compound composed of one carbon atom and four hydrogen atoms. Millions of years ago, the remains of plants and animals (diatoms) decayed and built up in thick layers. Scientists call this decayed matter from plants and animals organic material, since it was once alive. Over time, sand and silt settled and changed to rock, covered the organic material, and trapped it beneath the rock. Pressure and heat changed some of this organic material into coal, some into oil (petroleum), and some into natural gas—tiny bubbles of odorless gas. In some places, gas escapes from small gaps in rocks into the air. Then, if there is enough activation energy from lightning or a fire, it burns. When people first saw the flames, they experimented with them and learned they could use them for heat and light.

In about 200 B.C.E., the Chinese used natural gas to assist in the production of salt from salt brine. French explorers traveling in North America in 1616 encoun-

Characteristic light blue flame of a burner on a natural gas–fueled stove *(DOE/PNNL)*

tered Native Americans who were burning natural gas as it seeped into and around Lake Erie. The city of Baltimore used natural gas for street lights as early as 1816. An American businessman, William Hart, dug the first successful well intended to produce natural gas in 1821 in the town of Fredonia, New York. In 1859, another American businessman, Edwin Drake, drilled the first oil well in the United States in the village of Titusville, Pennsylvania. He discovered both petroleum and natural gas at a depth of just 69 feet (21 m) below the surface. Science historians generally regard Drake's well as the start of the American petroleum and natural gas industries.

Today, the search for natural gas begins with geologists, who study the structure and processes of the Earth. They locate the types of rock that are likely to contain gas and oil deposits. The tools used by geologists include seismic surveys that are used to find the right places to drill wells. Seismic surveys use echoes from a vibration source at the Earth's surface (usually a vibrating pad under a truck built for this purpose) to collect information about the rocks beneath. Sometimes it is necessary to use small amounts of dynamite to provide the vibration needed.

Scientists and engineers explore a chosen area by studying rock samples from deep below the ground and taking measurements. If the site seems promising, drilling begins. Some of these areas are on land, but

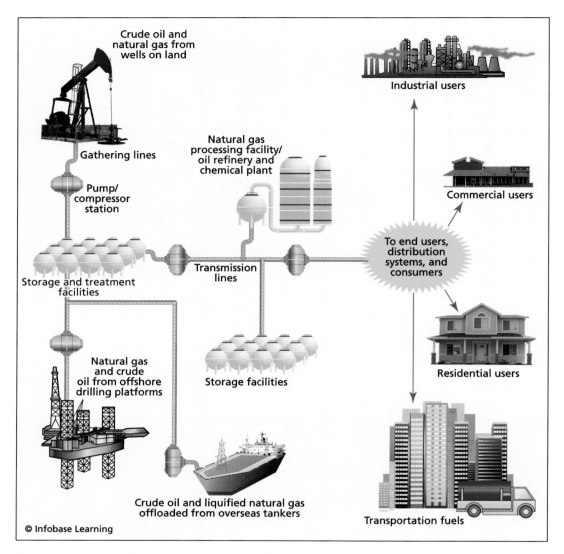

Crude oil and
natural gas from
wells on land

Gathering lines

Pump/
compressor
station

Storage and treatment
facilities

Natural gas
and crude
oil from offshore
drilling platforms

Natural gas
processing facility/
oil refinery and
chemical plant

Transmission
lines

Storage facilities

Crude oil and liquified natural gas
offloaded from overseas tankers

© Infobase Learning

Industrial users

Commercial users

To end users,
distribution
systems, and
consumers

Residential users

Transportation fuels

This figure depicts the major components of the natural gas and petroleum transmission pipeline system in the United States. *(DOT)*

many are offshore, deep in the ocean. Once the gas is found, it flows up through the well to the surface and into large pipelines.

Some of the gases that are produced along with methane, such as butane (C_4H_{10}) and propane (also known as by-products), are separated and cleaned at gas processing plants. The by-products, once removed, are used in a number of ways. For example, propane can be used for cooking on gas grills and to operate portable electric power generators. Dry natural gas is called consumer-grade natural gas. In addition to natural gas production, the U.S. gas supply is augmented by imports, withdrawals from storage, and supplemental gaseous fuels. Most of the natural gas consumed in the United States is produced in the United States. Some is imported from Canada and shipped to the United States in pipelines. Increasingly, natural gas is also being shipped to the United States as liquefied natural gas (LNG). People also use machines called "digesters" that turn today's organic materials (plants, animal wastes, etc.) into natural gas. This process replaces waiting millions of years for the gas to form naturally.

Natural gas is moved by pipelines from the producing fields to consumers. Because natural gas demand is greater in winter, it is stored along the way in large underground storage systems, such as old oil and gas wells and caverns formed in old salt beds. The gas remains there until it is added back into the pipeline when people begin to use more gas.

When the gas gets to the communities where it will be used, usually through large pipelines, it flows into smaller pipelines that engineers call mains. Very small lines, called services, connect to the mains and go directly to homes and buildings where it will be used.

The liquid natural gas (LNG) carrier *Khannur* *(NOAA)*

When chilled to very cold temperatures, approximately –260°F (–162°C), natural gas changes into a liquid and can be stored in this form. Because it takes up only 1/600th of the space that it would in its gaseous state, liquefied natural gas (LNG) can be loaded onto tankers (large ships with several domed tanks) and moved across oceans to other countries. When this LNG is received in the United States, it can be shipped by truck and be held in large chilled tanks close to users or turned back into gas when ready to be put in pipelines.

Liquefying natural gas provides a means of moving it long distances where pipeline transport is not feasible, allowing access to natural gas from regions with vast production potential that are too distant from end-use markets to be connected by pipeline. Representing a possible vision of tomorrow's surface transportation system in many parts of the United States, experimental vehicles, including buses, are being operated on either compressed natural gas (CNG) or liquid natural gas fuel.

According to the Department of Energy, about 25 percent of energy used in the United States in 2009 came from natural gas. That year, Americans consumed 22.84 trillion cubic feet (Tcf) (about 0.65 trillion m³) of

The compressed natural gas–fueled bus (left), liquid natural gas–fueled bus (center), and electric hybrid bus (right) used to transport visitors along the south rim area of Grand Canyon National Park *(NPS)*

natural gas. Natural gas is used to produce steel, glass, paper, clothing, bricks, and electricity and as an essential raw material for many common products. Some products that use natural gas as a raw material are paints, fertilizers, plastics, antifreeze, dyes, photographic film, medicines, and explosives. Slightly more than half of the homes in the United States use natural gas as their main heating fuel. Natural gas is also used in homes to fuel stoves, water heaters, clothes dryers, and other household appliances.

Natural gas burns more cleanly than other fossil fuels. It has fewer emissions of sulfur, carbon, and nitrogen than coal or oil, and when it is burned, it leaves almost no ash particles. Being a cleaner fuel, this is one reason that the use of natural gas, especially for electricity generation, has grown so much. However, as with other fossil fuels, burning natural gas produces carbon dioxide, which is a greenhouse gas.

As with other fuels, natural gas also affects the environment when it is produced, stored, and transported. Because natural gas is made up mostly of methane (yet another greenhouse gas), small amounts of methane can sometimes leak into the atmosphere from wells, storage tanks, and pipelines. The natural gas industry is working to prevent any methane from escaping. Exploring and drilling for natural gas will always have some impact on land and marine habitats, but new technologies have greatly reduced the number and size of areas disturbed by drilling, sometimes called "footprints." Furthermore, engineers now use horizontal and directional drilling techniques, which make it possible for a single well to produce gas from much bigger areas than in the past.

Natural gas pipelines and storage facilities have a good safety record. This is important because when natural gas leaks it can cause explosions. Since raw natural gas has no odor, natural gas companies add a distinctive, smelly substance to it so that people will know if there is a leak. A person with a natural gas stove may have smelled this "rotten egg" smell of natural gas when the pilot light has gone out.

METHANE HYDRATES

Some scientists believe that methane hydrates, ice that burns, will play an important role in the future global energy infrastructure. A gas hydrate is a crystalline solid; its building blocks consist of a gas molecule surrounded by a cage of water molecules. Therefore, it is similar to ice, except that the crystalline structure is stabilized by the "guest" gas molecule within the cage of water molecules. Many gases have molecular sizes suitable to form hydrates, including such naturally occurring gases as carbon dioxide, hydrogen sulfide, and several low-carbon-number hydrocarbons, but most marine gas hydrates that have been analyzed are methane hydrates.

Gas hydrates are a class of materials that the British scientist Sir Humphrey Davy first described in the early 1800s. They since have been described as an icelike crystalline mineral in which hydrocarbon gases and nonhydrocarbon gases are held within rigid cages of water molecules. Characterized in this way, gas hydrates may not sound very interesting or important, but they are. If a person holds a methane hydrate nodule in his or her hand and lights it with a match, the chunk will burn like a lantern wick.

Methane hydrate is a cagelike lattice of ice inside of which are trapped molecules of methane, the chief constituent of natural gas. If methane hydrate is either warmed or depressurized, it will revert back to water and natural gas. When brought to Earth's surface from the ocean bottom, 33.7 cubic feet (1 m³) of gas hydrate releases 5,730 ft³ (164 m³) of natural gas. Hydrate deposits may be several hundred feet (m) thick and generally occur in two types of settings: under Arctic permafrost, and beneath the ocean floor. Methane that forms hydrate can be both biogenic, created by biological activity in sediments, and thermogenic, created by geological processes deep within the Earth.

In the past decade, ocean seismic surveys, deep-sea drilling, and submersible studies have provided the first hints of the amount of gas hydrates that exist on Earth. The concentration of gas molecules locked into hydrate crystals is up to 160 times greater than the same volume of pure gas. The result may be the largest single reservoir of carbon on the planet—at least twice as much as all other fossil fuels (e.g., oil, natural gas, and coal) combined. The U.S. Exclusive Economic Zone (which extends to 200 nautical miles [370 km] offshore) may contain as much as 200,000 trillion cubic feet (5,663 trillion m³) of methane in hydrates. This is enough clean natural gas to power the United States for centuries.

Methane hydrate is known as the ice that burns. *(DOE/PNNL)*

While global estimates vary considerably, the energy content of methane that occurs in hydrate form is immense, possibly exceeding the combined energy content of all other known fossil fuels. However, future production volumes are speculative because methane production from hydrate has not been documented beyond small-scale field experiments.

Locating hydrates is one challenge; retrieving them is another. Mining them is not as easy as digging for coal or drilling for liquid. Several countries are now developing techniques to mine the gas locked inside hydrates as well as other fuels that may be trapped beneath the ice layer. For the last several years, laboratory scientists have been working on ways to commercially produce methane hydrates. The separation of methane from water molecules is relatively simple, but it is necessary to increase the temperature, decrease the pressure, or alter the reservoir's chemistry by introducing antifreeze. Nevertheless, many questions remain regarding this process of hydrate "dissociation" and the development of safe, economic means of gleaning methane from hydrates.

Based on some scientific studies, hydrates may join the legion of natural forces, such as orbital wobbles, mass volcanic eruptions, and ocean circulation, that drive global climate changes. Methane, the main type of hydrocarbon in hydrates, is a powerful greenhouse gas. It traps heat as carbon dioxide, the most well-known and abundant airborne greenhouse gas. Past massive meltdowns of subsea hydrates may, for example, have released enough methane to drive up global temperatures by 9° to 18°F (5° to 10°C) during the Late Paleocene era—some 55 million years ago. Such an increase in temperature today would be enough to melt the ice caps, raise sea levels by several feet (m), and transform America's wheat belt into a desert.

In a final apocalyptic note, below 1,640 feet (500 m) in temperate and subtropical oceans, such as those found off the continental United States, hydrate beds may lie just beneath or above the seafloor over much of the continental slope. Geologists speculate that massive submarine slumps, which can be likened to seafloor avalanches, may occur when hydrates break away from the steep slope. Such massive slumps may drive tidal waves that could drown miles of coastal shorelines. Only future research and field investigations will tell whether methane hydrates are a beneficial answer to global energy needs or a dangerous force of nature capable of causing major climate changes that could destroy modern civilization.

HYDROGEN

Scientists had been producing hydrogen for years before it was recognized as an element. Written records suggest that Robert Boyle produced hydrogen gas as early as 1671 while he was experimenting with iron and acids. Hydrogen was first recognized as a distinct element in 1866 by the British scientist Henry Cavendish (1731–1810) and later named by the French chemist Antoine Lavoisier. Starting in 1783, Lavoisier conducted a series of brilliant combustion experiments during which he demonstrated that water was composed of only hydrogen and oxygen.

Composed of a single proton and a single electron, hydrogen is the simplest and most abundant element in the universe. It is estimated that 90 percent of the visible universe is composed of hydrogen. Hydrogen is the basic thermonuclear fuel that most stars "burn" to release energy. The same process, known as fusion, is being studied as a possible power source for use on Earth. Astronomers anticipate that the Sun's supply of hydrogen will last another 5 billion years.

Hydrogen has three common isotopes. The simplest isotope, called protium, is ordinary hydrogen. The second, a stable isotope called deuterium, was discovered in 1932. The third isotope, called tritium, was discovered in 1934. Tritium is radioactive and has a half-life of 12.5 years.

Although hydrogen is the most abundant element in the universe, it is not found as a free element on Earth. Hydrogen gas is so much lighter than air that it rises and is quickly ejected from the atmosphere. This is why hydrogen as a gas (H_2) is not found by itself on Earth, but only in compound forms with other elements. Hydrogen combined with oxygen is water (H_2O). Hydrogen combined with carbon forms different compounds, including methane (CH_4), coal, and petroleum. Hydrogen is also found in all organic things—for example, biomass. It is also an abundant element in the Earth's crust.

The name hydrogen means "water forming." This gaseous element has an atomic number of 1, an atomic weight of 1.00794, a melting point of −434.81°F (−259.34°C [13.81 K]), and a boiling point of −423.1781°F (−252.87°C [20.28 K]).

Hydrogen is a commercially important element. Large amounts of hydrogen are combined with nitrogen from air to produce ammonia (NH_3) through a process called the Haber process. Hydrogen is also added to fats and oils, such as peanut oil, through a process called hydrogenation. Hydrogen combines with other elements to form numerous compounds. Some of the common ones are water (H_2O), ammonia (NH_3), methane (CH$_4$), table sugar (C$_{12}$H$_{22}$O$_{11}$), hydrogen peroxide (H$_2$O$_2$), and hydrochloric acid (HCl).

Liquid hydrogen is used in the study of superconductors and, when combined with liquid oxygen, makes an excellent rocket fuel. Liquid hydrogen also plays a role in the study and application of high-temperature superconductors. Energy engineers regard liquid hydrogen as a potential important energy carrier for the future.

An energy carrier is a substance or system that moves energy in a usable form from one place to another. Today, electricity is the most well known energy carrier. People use electricity to transport the energy available in coal, uranium, or falling water from generation facilities to points of application such as industrial sites and

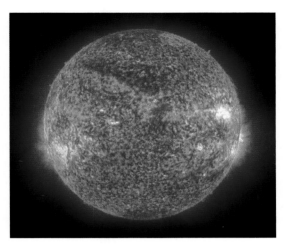

Hydrogen powers the visible universe. The Sun and all other main sequence stars obtain their energy by converting hydrogen into helium through thermonuclear reactions in their intensely hot cores. *(NASA/ESA)*

homes. An energy carrier makes the application of the energy content of the primary source more convenient. Liquid hydrogen can store energy until it is needed and can also move energy to a variety of places where it is needed.

Since hydrogen is not found as a free element on Earth, engineers have developed processes to separate it from other elements. At present, the two most common methods for producing hydrogen are steam reforming and electrolysis. The least expensive method is steam reforming, which accounts for about 95 percent of the hydrogen produced in the United States. In this process, hydrogen atoms are separated from the carbon atoms in methane (CH$_4$) molecules. Typically, chemical engineers employ an external source of hot gas to heat tubes in which a catalytic reaction takes place, converting steam (very hot H$_2$O) and lighter hydrocarbons into hydrogen and carbon monoxide (CO).

In the other production process, called electrolysis, hydrogen atoms are harvested after water molecules (H$_2$O) are split. Electrolysis is currently quite costly, but engineers are examining future approaches to hydrogen production at large central plants and small local plants. Once liquid hydrogen becomes less expensive and more available, it can be used in transportation systems as an environmentally friendly combustible fuel since the combustion by-product is water. Hydrogen can also be used in fuel cells to generate energy for vehicles, homes, and industrial sites.

As an extremely cold cryogen, proper equipment is needed to safely store and handle liquid hydrogen. Hydrogen gas is flammable, so care must also be exercised to avoid situations in which gas leaks could represent an explosive hazard. The flammable range of hydrogen in room temperature air at one atmosphere pressure is 4 to 75 percent by volume. The flame temperature of hydrogen in air is 3,713°F (2,045°C).

The American aerospace program has successfully handled liquid hydrogen for decades. The operational and safety procedures involving the space program's use of liquid hydrogen represent an important technical legacy. These experiences can assist the growth of a global hydrogen-based energy economy during this century.

Hydrogen fuel cells (batteries) make electricity. They are very efficient but expensive to build. Small fuel cells can power electric cars. Large fuel cells can provide electricity in remote places with no power lines. Because of the high cost to build fuel cells, large hydrogen power plants will probably not be constructed for some time. However, fuel cells are being used in some places as a source of emergency power, from hospitals to wilder-

The BMW Hydrogen 7 mono-fuel demonstration vehicle at the Argonne National Laboratory's Advanced Powertrain Research Facility *(DOE/ANL)*

ness locations. Portable fuel cells are being sold to provide longer power for laptop computers, cell phones, and military applications.

In 2010, there were an estimated 200 to 300 hydrogen-fueled vehicles in the United States. Most of these vehicles were buses and automobiles powered by electric motors. They store hydrogen gas or liquid onboard and convert the hydrogen into electricity for the motor using a fuel cell. Only a few of these vehicles burn the hydrogen directly, thereby producing almost no pollution.

The present cost of fuel cell vehicles greatly exceeds that of conventional vehicles, in large part due to the expense of producing fuel cells. However, hydrogen vehicles are starting to move from the laboratory to the road. Several are in use by a few state agencies and a few private entities. As of August 2010, there were 56 hydrogen refueling stations in the United States, about half of which were located in California. There are several so-called "chicken and egg" questions that hydrogen-powered vehicle developers are working hard to solve, including who will buy hydrogen cars if there are no refueling stations and who will pay to build a refueling station if there are no cars and customers. As these economic and technical issues are resolved over the next few decades, Earth's hydrogen-powered parent star will continue to shine upon the planet and make life possible.

CHAPTER 11

Conclusion

Everyone alive today and every person who has ever lived on Earth have shared at least one thing—the atmosphere. The average person takes some thousands of breaths a day. Out of the billions upon billions of air molecules inhaled in each breath, at least a few were also shared by some of history's most famous and infamous people, including Moses, Julius Caesar, Attila the Hun, Queen Elizabeth I of England, Abraham Lincoln, and John Wilkes Booth.

From a cosmic perspective, Earth is a spaceship hurling through the universe, and all crew members must share the same supply of oxygen found in the atmosphere. This sobering thought emphasizes the need for responsible stewardship of the planet. In the 21st century, humans must finally learn to appreciate the atmosphere and treat this protective gaseous envelope with more care and respect. Should the human family continue to allow irresponsible air pollution, people must brace themselves for the physical consequences of such foolhardy and self-destructive behavior.

Flighty and nomadic, gases nevertheless occupy a dominant position in the hierarchy of matter. At one time, all the matter in the universe consisted of the two gases hydrogen and helium, with perhaps a dash of lithium. One gas above all, air, dominates Earth's ability to host an incredible treasury of living creatures. Earth's thin atmospheric envelope protects and nurtures life but, quite paradoxically, also gives rise to enormously violent and destructive storms.

While discussing either weather or climate, scientists recognize that the atmosphere links all the major components of the Earth system (atmosphere, hydrosphere, cryosphere, biosphere, and solid Earth). Because weather influences people's daily activities, meteorologists attempt to make their weather forecasts accurate out to several days, typically a week, but they face significant challenges when attempting to predict the occurrence and behavior of adverse weather phenomena such as thunderstorms, tornados, and hurricanes. Climatologists take a much longer view and construct complex models of the Earth system in their efforts to predict changes in climate for a particular region or the entire planet. Some scientists, called paleoclimatologists, look far back into Earth's past in their efforts to develop better climate modeling tools.

The differences in atmospheric and surface temperatures from the equator to the poles set in motion a continuous movement of air and water around the planet. Great churning currents in the oceans and swirling winds in the atmosphere continuously carry heat from Earth's warmer equatorial regions toward the colder polar regions. The rising levels of carbon dioxide (CO_2) may herald a pending environmental calamity, due at least in part to human activities.

Fresh air is essential for life on Earth's surface, yet the World Health Organization (WHO) estimates that air pollution kills more than 8,000 people a day worldwide. There are natural sources of air pollution as well as anthropogenic (human-generated) sources. As discussed earlier, volcanic eruptions are not only spectacular displays of a restless planet but are a significant source of natural air pollution. Today, human beings pollute the air with vehicle emissions, gaseous discharges from industrial plants, and by combusting fossil fuels in large quantities to generate electricity. The naturally occurring radioactive gas radon is often overlooked but is a silent killer.

Gases played a major role in allowing scientists to develop an ability to relate the microscopic (atomic level) behavior of matter to readily observable macroscopic properties (such as density, pressure, and temperature). This capability transformed science and engineering and enabled the Scientific Revolution.

During the Scientific Revolution, pioneering scientists such as Galileo Galilei, Sir Isaac Newton, Blaise Pascal, Daniel Bernoulli, and Robert Boyle began describing and predicting fluid behavior. Their technical efforts yielded important fluid science relationships for both liquids and gases. They employed interesting experiments and mathematical relationships to help unlock nature's secrets. Their pioneering activities formed the broad field of fluid science.

Fluid science principles and concepts govern many fields, including astronomy, biology, medical sciences, chemistry, Earth science, meteorology, geology, oceanography, and physics. Fluid science forms an integral part of many branches of engineering, including aeronautics and astronautics, biological engineering, chemical engineering (including petroleum extraction and refining), civil engineering (including dam construction and calculating wind loads on tall buildings), environmental and sanitary engineering, industrial engineering (including food and beverage processing), mechanical engineering, nuclear engineering, and ocean engineering (including shipbuilding).

The modern aviation industry is one of humankind's greatest engineering achievements. Compressible gas dynamics is an integral part of aeronautics. Natural gas is an essential energy source in the modern world. Gases such as helium are essential in science and technology.

The American aerospace program has successfully handled hydrogen for decades. Operational and safety procedures involving the space program's use of liquid hydrogen represent an important technical legacy. These experiences can assist the growth of a global hydrogen-based energy economy this century.

Control of hydrogen, nature's most abundant element, represents the key to humankind's future on this planet and beyond. Hydrogen in fuel cells as the fuel for traditional thermodynamic power conversion appears to represent an environmentally friendly approach to electric power generation and transport. Harnessing controlled thermonuclear fusion offers major possibilities on Earth and makes the solar system available to more intense exploration and even human habitation. About 5 billion years ago, this solar system evolved from a huge cloud of hydrogen gas; now that same incredibly interesting gas holds the key to the future of the human race.

Appendix

Scientists correlate the properties of the elements portrayed in the periodic table with their electron configurations. Since, in a neutral atom, the number of electrons equals the number of protons, they arrange the elements in order of their increasing atomic number (Z). The modern periodic table has seven horizontal rows (called periods) and 18 vertical columns (called groups). The properties of the elements in a particular row vary across it, providing the concept of periodicity.

There are several versions of the periodic table used in modern science. The International Union of Pure and Applied Chemistry (IUPAC) recommends labeling the vertical columns from 1 to 18, starting with hydrogen (H) as the top of group 1 and ending with helium (He) as the top of group 18. The IUPAC further recommends labeling the periods (rows) from 1

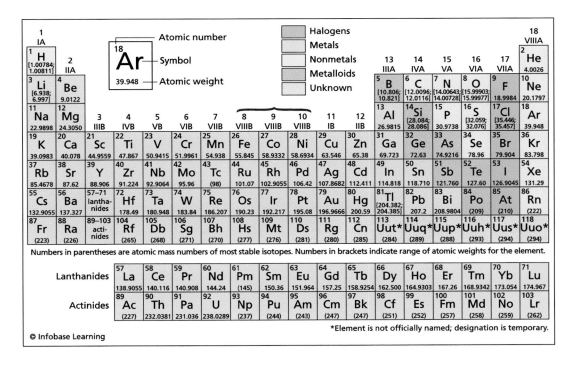

© Infobase Learning

to 7. Hydrogen (H) and helium (He) are the only two elements found in period (row) 1. Period 7 starts with francium (Fr) and includes the actinide series as well as the transactinides (very short-lived, human-made, super-heavy elements).

The row (or period) in which an element appears in the periodic table tells scientists how many electron shells an atom of that particular element possesses. The column (or group) lets scientists know how many electrons to expect in an element's outermost electron shell. Scientists call an electron residing in an atom's outermost shell a valence electron. Chemists have learned that it is these valence electrons that determine the chemistry of a particular element. The periodic table is structured such that all the elements in the same column (group) have the same number of valence electrons. The elements that appear in a particular column (group) display similar chemistry.

ELEMENTS LISTED BY ATOMIC NUMBER

1	H	Hydrogen	18	Ar	Argon
2	He	Helium	19	K	Potassium
3	Li	Lithium	20	Ca	Calcium
4	Be	Beryllium	21	Sc	Scandium
5	B	Boron	22	Ti	Titanium
6	C	Carbon	23	V	Vanadium
7	N	Nitrogen	24	Cr	Chromium
8	O	Oxygen	25	Mn	Manganese
9	F	Fluorine	26	Fe	Iron
10	Ne	Neon	27	Co	Cobalt
11	Na	Sodium	28	Ni	Nickel
12	Mg	Magnesium	29	Cu	Copper
13	Al	Aluminum	30	Zn	Zinc
14	Si	Silicon	31	Ga	Gallium
15	P	Phosphorus	32	Ge	Germanium
16	S	Sulfur	33	As	Arsenic
17	Cl	Chlorine	34	Se	Selenium

35	Br	Bromine	64	Gd	Gadolinium
36	Kr	Krypton	65	Tb	Terbium
37	Rb	Rubidium	66	Dy	Dysprosium
38	Sr	Strontium	67	Ho	Holmium
39	Y	Yttrium	68	Er	Erbium
40	Zr	Zirconium	69	Tm	Thulium
41	Nb	Niobium	70	Yb	Ytterbium
42	Mo	Molybdenum	71	Lu	Lutetium
43	Tc	Technetium	72	Hf	Hafnium
44	Ru	Ruthenium	73	Ta	Tantalum
45	Rh	Rhodium	74	W	Tungsten
46	Pd	Palladium	75	Re	Rhenium
47	Ag	Silver	76	Os	Osmium
48	Cd	Cadmium	77	Ir	Iridium
49	In	Indium	78	Pt	Platinum
50	Sn	Tin	79	Au	Gold
51	Sb	Antimony	80	Hg	Mercury
52	Te	Tellurium	81	Tl	Thallium
53	I	Iodine	82	Pb	Lead
54	Xe	Xenon	83	Bi	Bismuth
55	Cs	Cesium	84	Po	Polonium
56	Ba	Barium	85	At	Astatine
57	La	Lanthanum	86	Rn	Radon
58	Ce	Cerium	87	Fr	Francium
59	Pr	Praseodymium	88	Ra	Radium
60	Nd	Neodymium	89	Ac	Actinium
61	Pm	Promethium	90	Th	Thorium
62	Sm	Samarium	91	Pa	Protactinium
63	Eu	Europium	92	U	Uranium

(continues)

ELEMENTS LISTED BY ATOMIC NUMBER *(continued)*

93	Np	Neptunium		106	Sg	Seaborgium
94	Pu	Plutonium		107	Bh	Bohrium
95	Am	Americium		108	Hs	Hassium
96	Cm	Curium		109	Mt	Meitnerium
97	Bk	Berkelium		110	Ds	Darmstadtium
98	Cf	Californium		111	Rg	Roentgenium
99	Es	Einsteinium		112	Cn	Copernicum
100	Fm	Fermium		113	Uut	Ununtrium
101	Md	Mendelevium		114	Uuq	Ununquadium
102	No	Nobelium		115	Uup	Ununpentium
103	Lr	Lawrencium		116	Uuh	Ununhexium
104	Rf	Rutherfordium		117	Uus	Ununseptium
105	Db	Dubnium		118	Uuo	Ununoctium

Chronology

Civilization is essentially the story of the human mind understanding and gaining control over matter. The chronology presents some of the major milestones, scientific breakthroughs, and technical developments that formed the modern understanding of matter. Note that dates prior to 1543 are approximate.

13.7 BILLION YEARS AGO. . . . . Big bang event starts the universe.

13.3 BILLION YEARS AGO. . . . . The first stars form and begin to shine intensely.

4.5 BILLION YEARS AGO. . . . . . Earth forms within the primordial solar nebula.

3.6 BILLION YEARS AGO. . . . . . Life (simple microorganisms) appears in Earth's oceans.

2,000,000–100,000 B.C.E.. . . Early hunters of the Lower Paleolithic learn to use simple stone tools, such as handheld axes.

100,000–40,000 B.C.E. . . . . Neanderthal man of Middle Paleolithic lives in caves, controls fire, and uses improved stone tools for hunting.

40,000–10,000 B.C.E. . . . . . During the Upper Paleolithic, Cro-Magnon man displaces Neanderthal man. Cro-Magnon people develop more organized hunting and fishing activities using improved stone tools and weapons.

8000–3500 B.C.E.. . . . . Neolithic Revolution takes place in the ancient Middle East as people shift their dependence for subsistence from hunting and gathering to crop cultivation and animal domestication.

3500–1200 B.C.E.. . . . . Bronze Age occurs in the ancient Middle East, when metalworking artisans start using bronze (a copper and tin alloy) to make weapons and tools.

1200–600 B.C.E.. . . . . People in the ancient Middle East enter the Iron Age. Eventually, the best weapons and tools are made of steel, an alloy of iron and varying amounts of carbon. The improved metal tools and weapons spread to Greece and later to Rome.

1000 B.C.E. . . . . By this time, people in various ancient civilizations have discovered and are using the following chemical elements (in alphabetical order): carbon (C), copper (Cu), gold (Au), iron (Fe), lead (Pb), mercury (Hg), silver (Ag), sulfur (S), tin (Sn), and zinc (Zn).

650 B.C.E. . . . . Kingdom of Lydia introduces officially minted gold and silver coins.

600 B.C.E. . . . . Early Greek philosopher Thales of Miletus postulates that all substances come from water and would eventually turn back into water.

450 B.C.E. . . . . Greek philosopher Empedocles proposes that all matter is made up of four basic elements (earth, air, water, and fire) that periodically combine and separate under the influence of two opposing forces (love and strife).

430 B.C.E. . . . . Greek philosopher Democritus proposes that all things consist of changeless, indivisible, tiny pieces of matter called *atoms.*

250 B.C.E. . . . . Archimedes of Syracuse designs an endless screw, later called the Archimedes screw. People use the fluid-moving device to remove water from the holds of sailing ships and to irrigate arid fields.

300 C.E. . . . . Greek alchemist Zosimos of Panoplis writes the oldest known work describing alchemy.

850 . . . . The Chinese use gunpowder for festive fireworks. It is a mixture of sulfur (S), charcoal (C), and potassium nitrate (KNO_3).

1247 . . . . British monk Roger Bacon writes the formula for gunpowder in his encyclopedic work *Opus Majus.*

1250 . . . . German theologian and natural philosopher Albertus Magnus isolates the element arsenic (As).

1439 . . . . Johannes Gutenberg successfully incorporates movable metal type in his mechanical printing press. His revolutionary approach to printing depends on a durable, hard metal alloy called type metal, which consists of a mixture of lead (Pb), tin (Sn), and antimony (Sb).

1543 . . . . Start of the Scientific Revolution. Polish astronomer Nicholas Copernicus promotes heliocentric (Sun-centered) cosmology with his deathbed publication of *On the Revolutions of Celestial Orbs.*

1638 . . . . Italian scientist Galileo Galilei publishes extensive work on solid mechanics, including uniform acceleration, free fall, and projectile motion.

1643 . . . . Italian physicist Evangelista Torricelli designs the first mercury barometer and then records the daily variation of atmospheric pressure.

1661 . . . . Irish-British scientist Robert Boyle publishes *The Sceptical Chymist,* in which he abandons the four classical Greek elements (earth, air, water, and fire) and questions how alchemists determine what substances are elements.

1665 . . . . British scientist Robert Hooke publishes *Micrographia,* in which he describes pioneering applications of the optical microscope in chemistry, botany, and other scientific fields.

1667 . . . . The work of German alchemist Johann Joachim Becher forms the basis of the phlogiston theory of heat.

1669 . . . . German alchemist Hennig Brand discovers the element phosphorous (P).

1678 . . . . Robert Hooke studies the action of springs and reports that the extension (or compression) of an elastic material takes place in direct proportion to the force exerted on the material.

1687 . . . . British physicist Sir Isaac Newton publishes *The Principia.* His work provides the mathematical foundations for understanding (from a classical physics perspective) the motion of almost everything in the physical universe.

1738 .... Swiss mathematician Daniel Bernoulli publishes *Hydrodynamica*. In this seminal work, he identifies the relationships between density, pressure, and velocity in flowing fluids.

1748 .... While conducting experiments with electricity, American statesman and scientist Benjamin Franklin coins the term *battery*.

1754 .... Scottish chemist Joseph Black discovers a new gaseous substance, which he calls "fixed air." Other scientists later identify it as carbon dioxide (CO_2).

1764 .... Scottish engineer James Watt greatly improves the Newcomen steam engine. Watt steam engines power the First Industrial Revolution.

1772 .... Scottish physician and chemist Daniel Rutherford isolates a new colorless gaseous substance, calling it "noxious air." Other scientists soon refer to the new gas as nitrogen (N_2).

1785 .... French scientist Charles-Augustin de Coulomb performs experiments that lead to the important law of electrostatics, later known as Coulomb's law.

1789 .... French chemist Antoine-Laurent Lavoisier publishes *Treatise of Elementary Chemistry*, the first modern textbook on chemistry. Lavoisier also promotes the caloric theory of heat.

1800 .... Italian physicist Count Alessandro Volta invents the voltaic pile. His device is the forerunner of the modern electric battery.

1803 .... British schoolteacher and chemist John Dalton revives the atomic theory of matter. From his experiments, he concludes that all matter consists of combinations of atoms and that all the atoms of a particular element are identical.

1807 .... British chemist Sir Humphry Davy discovers the element potassium (K) while experimenting with caustic potash (KOH). Potassium is the first metal isolated by the process of electrolysis.

1811................... Italian physicist Amedeo Avogadro proposes that equal volumes of different gases under the same conditions of pressure and temperature contain the same number of molecules. Scientists call this important hypothesis Avogadro's law.

1820 .... Danish physicist Hans Christian Ørsted discovers a relationship between magnetism and electricity.

1824 .... French military engineer Sadi Carnot publishes *Reflections on the Motive Power of Fire.* Despite the use of caloric theory, his work correctly identifies the general thermodynamic principles that govern the operation and efficiency of all heat engines.

1826 .... French scientist André-Marie Ampère experimentally formulates the relationship between electricity and magnetism.

1827 .... Experiments performed by German physicist George Simon Ohm indicate a fundamental relationship among voltage, current, and resistance.

1828 .... Swedish chemist Jöns Jacob Berzelius discovers the element thorium (Th).

1831 .... British experimental scientist Michael Faraday discovers the principle of electromagnetic induction. This principle is the basis for the electric dynamo.

Independent of Faraday, the American physicist Joseph Henry publishes a paper describing the electric motor (essentially a reverse dynamo).

1841 .... German physicist and physician Julius Robert von Mayer states the conservation of energy principle, namely that energy can neither be created nor destroyed.

1847 .... British physicist James Prescott Joule experimentally determines the mechanical equivalent of heat. Joule's work is a major step in developing the modern science of thermodynamics.

1866 .... Swedish scientist-industrialist Alfred Nobel finds a way to stabilize nitroglycerin and calls the new chemical explosive mixture dynamite.

1869 .... Russian chemist Dmitri Mendeleev introduces a periodic listing of the 63 known chemical elements in *Principles of Chemistry*. His periodic table includes gaps for elements predicted but not yet discovered.

American printer John W. Hyatt formulates celluloid, a flammable thermoplastic material made from a mixture of cellulose nitrate, alcohol, and camphor.

1873 .... Scottish mathematician and theoretical physicist James Clerk Maxwell publishes *Treatise on Electricity and Magnetism*.

1876 .... American physicist and chemist Josiah Willard Gibbs publishes *On the Equilibrium of Heterogeneous Substances*. This compendium forms the theoretical foundation of physical chemistry.

1884 .... Swedish chemist Svante Arrhenius proposes that electrolytes split or dissociate into electrically opposite positive and negative ions.

1888 .... German physicist Heinrich Rudolf Hertz produces and detects radio waves.

1895 .... German physicist Wilhelm Conrad Roentgen discovers X-rays.

1896 .... While investigating the properties of uranium salt, French physicist Antoine-Henri Becquerel discovers radioactivity.

1897 .... British physicist Sir Joseph John Thomson performs experiments that demonstrate the existence of the electron—the first subatomic particle discovered.

1898 .... French scientists Pierre and (Polish-born) Marie Curie announce the discovery of two new radioactive elements, polonium (Po) and radium (Ra).

1900 .... German physicist Max Planck postulates that blackbodies radiate energy only in discrete packets (or quanta) rather than continuously. His hypothesis marks the birth of quantum theory.

1903 .... New Zealand–born British physicist Baron (Ernest) Rutherford and British radiochemist Frederick Soddy propose the law of radioactive decay.

1904 .... German physicist Ludwig Prandtl revolutionizes fluid mechanics by introducing the concept of the boundary layer and its role in fluid flow.

1905 .... Swiss-German-American physicist Albert Einstein publishes the special theory of relativity, including the famous mass-energy equivalence formula ($E = mc^2$).

1907 .... Belgian-American chemist Leo Baekeland formulates bakelite. This synthetic thermoplastic material ushers in the age of plastics.

1911 .... Baron Ernest Rutherford proposes the concept of the atomic nucleus based on the startling results of an alpha particle–gold foil scattering experiment.

1912 .... German physicist Max von Laue discovers that X-rays are diffracted by crystals.

1913 .... Danish physicist Niels Bohr presents his theoretical model of the hydrogen atom—a brilliant combination of atomic theory with quantum physics.

Frederick Soddy proposes the existence of isotopes.

1914 .... British physicist Henry Moseley measures the characteristic X-ray lines of many chemical elements.

1915 .... Albert Einstein presents his general theory of relativity, which relates gravity to the curvature of space-time.

1919 .... Ernest Rutherford bombards nitrogen (N) nuclei with alpha particles, causing the nitrogen nuclei to transform into oxygen (O) nuclei and to emit protons (hydrogen nuclei).

British physicist Francis Aston uses the newly invented mass spectrograph to identify more than 200 naturally occurring isotopes.

1923 .... American physicist Arthur Holly Compton conducts experiments involving X-ray scattering that demonstrate the particle nature of energetic photons.

1924 .... French physicist Louis-Victor de Broglie proposes the particle-wave duality of matter.

1926 .... Austrian physicist Erwin Schrödinger develops quantum wave mechanics to describe the dual wave-particle nature of matter.

1927 .... German physicist Werner Heisenberg introduces his uncertainty principle.

1929 .... American astronomer Edwin Hubble announces that his observations of distant galaxies suggest an expanding universe.

1932 .... British physicist Sir James Chadwick discovers the neutron.

British physicist Sir John Cockcroft and Irish physicist Ernest Walton use a linear accelerator to bombard lithium (Li) with energetic protons, producing the first artificial disintegration of an atomic nucleus.

American physicist Carl D. Anderson discovers the positron.

1934 .... Italian-American physicist Enrico Fermi proposes a theory of beta decay that includes the neutrino. He also starts to bombard uranium with neutrons and discovers the phenomenon of slow neutrons.

1938 .... German chemists Otto Hahn and Fritz Strassmann bombard uranium with neutrons and detect the presence of lighter elements. Austrian physicist Lise Meitner and Austrian-British physicist Otto Frisch review Hahn's work and conclude in early 1939 that the German chemists had split the atomic nucleus, achieving neutron-induced nuclear fission.

E.I. du Pont de Nemours & Company introduces a new thermoplastic material called nylon.

1941 .... American nuclear scientist Glenn T. Seaborg and his associates use the cyclotron at the University of California, Berkeley, to synthesize plutonium (Pu).

1942 .... Modern nuclear age begins when Enrico Fermi's scientific team at the University of Chicago achieves the first self-sustained, neutron-induced fission chain reaction at Chicago Pile One (CP-1), a uranium-fueled, graphite-moderated atomic pile (reactor).

1945 .... American scientists successfully detonate the world's first nuclear explosion, a plutonium-implosion device code-named Trinity.

1947 .... American physicists John Bardeen, Walter Brattain, and William Shockley invent the transistor.

1952 .... A consortium of 11 founding countries establishes CERN, the European Organization for Nuclear Research, at a site near Geneva, Switzerland.

United States tests the world's first thermonuclear device (hydrogen bomb) at the Enewetak Atoll in the Pacific Ocean. Code-named Ivy Mike, the experimental device produces a yield of 10.4 megatons.

1964 .... German-American physicist Arno Allen Penzias and American physicist Robert Woodrow Wilson detect the cosmic microwave background (CMB).

1967 .... German-American physicist Hans Albrecht Bethe receives the 1967 Nobel Prize in physics for his theory of thermonuclear reactions being responsible for energy generation in stars.

1969 .... On July 20, American astronauts Neil Armstrong and Edwin "Buzz" Aldrin successfully land on the Moon as part of NASA's *Apollo 11* mission.

1972 .... NASA launches the *Pioneer 10* spacecraft. It eventually becomes the first human-made object to leave the solar system on an interstellar trajectory

1985 .... American chemists Robert F. Curl, Jr., and Richard E. Smalley, collaborating with British astronomer

Sir Harold W. Kroto, discover the buckyball, an allotrope of pure carbon.

1996 .... Scientists at CERN (near Geneva, Switzerland) announce the creation of antihydrogen, the first human-made antimatter atom.

1998 .... Astrophysicists investigating very distant Type 1A supernovae discover that the universe is expanding at an accelerated rate. Scientists coin the term *dark energy* in their efforts to explain what these observations physically imply.

2001 .... American physicist Eric A. Cornell, German physicist Wolfgang Ketterle, and American physicist Carl E. Wieman share the 2001 Nobel Prize in physics for their fundamental studies of the properties of Bose-Einstein condensates.

2005 .... Scientists at the Lawrence Livermore National Laboratory (LLNL) in California and the Joint Institute for Nuclear Research (JINR) in Dubna, Russia, perform collaborative experiments that establish the existence of super-heavy element 118, provisionally called ununoctium (Uuo).

2008 .... An international team of scientists inaugurates the world's most powerful particle accelerator, the Large Hadron Collider (LHC), located at the CERN laboratory near Geneva, Switzerland.

2009 .... British scientist Charles Kao, American scientist Willard Boyle, and American scientist George Smith share the 2009 Nobel Prize in physics for their pioneering efforts in fiber optics and imaging semiconductor devices, developments that unleashed the information technology revolution.

2010 .... Element 112 is officially named Copernicum (Cn) by the IUPAC in honor of Polish astronomer Nicholas Copernicus (1473–1543), who championed heliocentric cosmology.

Scientists at the Joint Institute for Nuclear Research in Dubna, Russia, announce the synthesis of element 117 (ununseptium [Uus]) in early April.

Glossary

absolute zero   the lowest possible temperature; equal to 0 kelvin (K) (−459.67°F, −273.15°C)

acceleration (a)   rate at which the velocity of an object changes with time

accelerator   device for increasing the velocity and energy of charged elementary particles

acid   substance that produces hydrogen ions (H^+) when dissolved in water

actinoid (formerly actinide)   series of heavy metallic elements beginning with element 89 (actinium) and continuing through element 103 (lawrencium)

activity   measure of the rate at which a material emits nuclear radiations

air   overall mixture of gases that make up Earth's atmosphere

alchemy   mystical blend of sorcery, religion, and prescientific chemistry practiced in many early societies around the world

alloy   solid solution (compound) or homogeneous mixture of two or more elements, at least one of which is an elemental metal

alpha particle (α)   positively charged nuclear particle emitted from the nucleus of certain radioisotopes when they undergo decay; consists of two protons and two neutrons bound together

alternating current (AC)   electric current that changes direction periodically in a circuit

American customary system of units (also American system)   used primarily in the United States; based on the foot (ft), pound-mass (lbm), pound-force (lbf), and second (s). Peculiar to this system is the artificial construct (based on Newton's second law) that one pound-force equals one pound-mass (lbm) at sea level on Earth

ampere (A)   SI unit of electric current

anode   positive electrode in a battery, fuel cell, or electrolytic cell; oxidation occurs at anode

antimatter   matter in which the ordinary nuclear particles are replaced by corresponding antiparticles

Archimedes principle   the fluid mechanics rule that states that the buoyant (upward) force exerted on a solid object immersed in a fluid equals the weight of the fluid displaced by the object

atom   smallest part of an element, indivisible by chemical means; consists of a dense inner core (nucleus) that contains protons and neutrons and a cloud of orbiting electrons

atomic mass   *See* **relative atomic mass**

atomic mass unit (amu)   1/12 mass of carbon's most abundant isotope, namely carbon-12

atomic number (Z)   total number of protons in the nucleus of an atom and its positive charge

atomic weight   the mass of an atom relative to other atoms. *See also* **relative atomic mass**

battery   electrochemical energy storage device that serves as a source of direct current or voltage

becquerel (Bq)   SI unit of radioactivity; one disintegration (or spontaneous nuclear transformation) per second. *Compare with* **curie**

beta particle (β)   elementary particle emitted from the nucleus during radioactive decay; a negatively charged beta particle is identical to an electron

big bang   theory in cosmology concerning the origin of the universe; postulates that about 13.7 billion years ago, an initial singularity experienced a very large explosion that started space and time. Astrophysical observations support this theory and suggest that the universe has been expanding at different rates under the influence of gravity, dark matter, and dark energy

blackbody   perfect emitter and perfect absorber of electromagnetic radiation; radiant energy emitted by a blackbody is a function only of the emitting object's absolute temperature

black hole   incredibly compact, gravitationally collapsed mass from which nothing can escape

boiling point   temperature (at a specified pressure) at which a liquid experiences a change of state into a gas

Bose-Einstein condensate (BEC)   state of matter in which extremely cold atoms attain the same quantum state and behave essentially as a large "super atom"

boson general name given to any particle with a spin of an integral number (0, 1, 2, etc.) of quantum units of angular momentum. Carrier particles of all interactions are bosons. *See also* **carrier particle**

brass alloy of copper (Cu) and zinc (Zn)

British thermal unit (Btu) amount of heat needed to raise the temperature of 1 lbm of water 1°F at normal atmospheric pressure; 1 Btu = 1,055 J = 252 cal

bronze alloy of copper (Cu) and tin (Sn)

calorie (cal) quantity of heat; defined as the amount needed to raise one gram of water 1°C at normal atmospheric pressure; 1 cal = 4.1868 J = 0.004 Btu

carbon dioxide (CO_2) colorless, odorless, noncombustible gas present in Earth's atmosphere

Carnot cycle ideal reversible thermodynamic cycle for a theoretical heat engine; represents the best possible thermal efficiency of any heat engine operating between two absolute temperatures (T_1 and T_2)

carrier particle within the standard model, gluons are carrier particles for strong interactions; photons are carrier particles of electromagnetic interactions; and the W and Z bosons are carrier particles for weak interactions. *See also* **standard model**

catalyst substance that changes the rate of a chemical reaction without being consumed or changed by the reaction

cathode negative electrode in a battery, fuel cell, electrolytic cell, or electron (discharge) tube through which a primary stream of electrons enters a system

chain reaction reaction that stimulates its own repetition. *See also* **nuclear chain reaction**

change of state the change of a substance from one physical state to another; the atoms or molecules are structurally rearranged without experiencing a change in composition. Sometimes called change of phase or phase transition

charged particle elementary particle that carries a positive or negative electric charge

chemical bond(s) force(s) that holds atoms together to form stable configurations of molecules

chemical property characteristic of a substance that describes the manner in which the substance will undergo a reaction with another substance, resulting in a change in chemical composition. *Compare with* **physical property**

chemical reaction involves changes in the electron structure surrounding the nucleus of an atom; a dissociation, recombination, or rearrangement of atoms. During a chemical reaction, one or more kinds of matter (called reactants) are transformed into one or several new kinds of matter (called products)

color charge in the standard model, the charge associated with strong interactions. Quarks and gluons have color charge and thus participate in strong interactions. Leptons, photons, W bosons, and Z bosons do not have color charge and consequently do not participate in strong interactions. *See also* **standard model**

combustion chemical reaction (burning or rapid oxidation) between a fuel and oxygen that generates heat and usually light

composite materials human-made materials that combine desirable properties of several materials to achieve an improved substance; includes combinations of metals, ceramics, and plastics with built-in strengthening agents

compound pure substance made up of two or more elements chemically combined in fixed proportions

compressible flow fluid flow in which density changes cannot be neglected

compression condition when an applied external force squeezes the atoms of a material closer together. *Compare* **tension**

concentration for a solution, the quantity of dissolved substance per unit quantity of solvent

condensation change of state process by which a vapor (gas) becomes a liquid. *The opposite of* **evaporation**

conduction (thermal) transport of heat through an object by means of a temperature difference from a region of higher temperature to a region of lower temperature. *Compare with* **convection**

conservation of mass and energy Einstein's special relativity principle stating that energy (E) and mass (m) can neither be created nor destroyed, but are interchangeable in accordance with the equation $E = mc^2$, where c represents the speed of light

convection fundamental form of heat transfer characterized by mass motions within a fluid resulting in the transport and mixing of the properties of that fluid

coulomb (C)   SI unit of electric charge; equivalent to quantity of electric charge transported in one second by a current of one ampere

covalent bond   the chemical bond created within a molecule when two or more atoms share an electron

creep   slow, continuous, permanent deformation of solid material caused by a constant tensile or compressive load that is less than the load necessary for the material to give way (yield) under pressure. *See also* **plastic deformation**

crystal   a solid whose atoms are arranged in an orderly manner, forming a distinct, repetitive pattern

curie (Ci)   traditional unit of radioactivity equal to 37 billion (37×10^9) disintegrations per second. *Compare with* **becquerel**

current (I)   flow of electric charge through a conductor

dark energy   a mysterious natural phenomenon or unknown cosmic force thought responsible for the observed acceleration in the rate of expansion of the universe. Astronomical observations suggest dark energy makes up about 72 percent of the universe

dark matter (nonbaryonic matter)   exotic form of matter that emits very little or no electromagnetic radiation. It experiences no measurable interaction with ordinary (baryonic) matter but somehow accounts for the observed structure of the universe. It makes up about 23 percent of the content of the universe, while ordinary matter makes up less than 5 percent

density (ρ)   mass of a substance per unit volume at a specified temperature

deposition   direct transition of a material from the gaseous (vapor) state to the solid state without passing through the liquid phase. *Compare* with **sublimation**

dipole magnet   any magnet with one north and one south pole

direct current (DC)   electric current that always flows in the same direction through a circuit

elastic deformation   temporary change in size or shape of a solid due to an applied force (stress); when force is removed the solid returns to its original size and shape

elasticity   ability of a body that has been deformed by an applied force to return to its original shape when the force is removed

elastic modulus   a measure of the stiffness of a solid material; defined as the ratio of stress to strain

electricity   flow of energy due to the motion of electric charges; any physical effect that results from the existence of moving or stationary electric charges

electrode   conductor (terminal) at which electricity passes from one medium into another; positive electrode is the *anode;* negative electrode is the *cathode*

electrolyte   a chemical compound that, in an aqueous (water) solution, conducts an electric current

electromagnetic radiation (EMR)   oscillating electric and magnetic fields that propagate at the speed of light. Includes in order of increasing frequency and energy: radio waves, radar waves, infrared (IR) radiation, visible light, ultraviolet radiation, X-rays, and gamma rays

electron (e)   stable elementary particle with a unit negative electric charge (1.602×10^{-19} C). Electrons form an orbiting cloud, or shell, around the positively charged atomic nucleus and determine an atom's chemical properties

electron volt (eV)   energy gained by an electron as it passes through a potential difference of one volt; one electron volt has an energy equivalence of 1.519×10^{-22} Btu $= 1.602 \times 10^{-19}$ J

element   pure chemical substance indivisible into simpler substances by chemical means; all the atoms of an element have the same number of protons in the nucleus and the same number of orbiting electrons, although the number of neutrons in the nucleus may vary

elementary particle   a fundamental constituent of matter; the basic atomic model suggests three elementary particles: the proton, neutron, and electron. *See also* **fundamental particle**

endothermic reaction   chemical reaction requiring an input of energy to take place. *Compare* **exothermic reaction**

energy (E)   capacity to do work; appears in many different forms, such as mechanical, thermal, electrical, chemical, and nuclear

entropy (S)   measure of disorder within a system; as entropy increases, energy becomes less available to perform useful work

evaporation   physical process by which a liquid is transformed into a gas (vapor) at a temperature below the boiling point of the liquid. *Compare with* **sublimation**

excited state   state of a molecule, atom, electron, or nucleus when it possesses more than its normal energy. *Compare with* **ground state**

exothermic reaction   chemical reaction that releases energy as it takes place. *Compare with* **endothermic reaction**

fatigue   weakening or deterioration of metal or other material that occurs under load, especially under repeated cyclic or continued loading

fermion   general name scientists give to a particle that is a matter constituent. Fermions are characterized by spin in odd half-integer quantum units (namely, 1/2, 3/2, 5/2, etc.); quarks, leptons, and baryons are all fermions

fission (nuclear)   splitting of the nucleus of a heavy atom into two lighter nuclei accompanied by the release of a large amount of energy as well as neutrons, X-rays, and gamma rays

flavor   in the standard model, quantum number that distinguishes different types of quarks and leptons. *See also* **quark; lepton**

fluid mechanics   scientific discipline that deals with the behavior of fluids (both gases and liquids) at rest (fluid statics) and in motion (fluid dynamics)

foot-pound (force) (ft-lb$_{force}$)   unit of work in American customary system of units; 1 ft-lb$_{force}$ = 1.3558 J

force (F)   the cause of the acceleration of material objects as measured by the rate of change of momentum produced on a free body. Force is a vector quantity mathematically expressed by Newton's second law of motion: force = mass × acceleration

freezing point   the temperature at which a substance experiences a change from the liquid state to the solid state at a specified pressure; at this temperature, the solid and liquid states of a substance can coexist in equilibrium. *Synonymous with* **melting point**

fundamental particle   particle with no internal substructure; in the standard model, any of the six types of quarks or six types of leptons and their antiparticles. Scientists postulate that all other particles are made from a combination of quarks and leptons. *See also* **elementary particle**

fusion (nuclear)   nuclear reaction in which lighter atomic nuclei join together (fuse) to form a heavier nucleus, liberating a great deal of energy

g   acceleration due to gravity at sea level on Earth; approximately 32.2 ft/s^2 (9.8 m/s^2)

gamma ray (γ)   high-energy, very short–wavelength photon of electromagnetic radiation emitted by a nucleus during certain nuclear reactions or radioactive decay

gas   state of matter characterized as an easily compressible fluid that has neither a constant volume nor a fixed shape; a gas assumes the total size and shape of its container

gravitational lensing   bending of light from a distant celestial object by a massive (gravitationally influential) foreground object

ground state   state of a nucleus, atom, or molecule at its lowest (normal) energy level

hadron   any particle (such as a baryon) that exists within the nucleus of an atom; made up of quarks and gluons, hadrons interact with the strong force

half-life (radiological)   time in which half the atoms of a particular radioactive isotope disintegrate to another nuclear form

heat   energy transferred by a temperature difference or thermal process. *Compare* **work**

heat capacity (c)   amount of heat needed to raise the temperature of an object by one degree

heat engine   thermodynamic system that receives energy in the form of heat and that, in the performance of energy transformation on a working fluid, does work. Heat engines function in thermodynamic cycles

hertz (Hz)   SI unit of frequency; equal to one cycle per second

high explosive (HE)   energetic material that detonates (rather than burns); the rate of advance of the reaction zone into the unreacted material exceeds the velocity of sound in the unreacted material

horsepower (hp)   American customary system unit of power; 1 hp = 550 ft-lb$_{force}$/s = 746 W

hydraulic   operated, moved, or affected by liquid used to transmit energy

hydrocarbon   organic compound composed of only carbon and hydrogen atoms

ideal fluid   *See* **perfect fluid**

ideal gas law   important physical principle: $P V = n R_u T$, where P is pressure, V is volume, T is temperature, n is the number of moles of gas, and R_u is the universal gas constant

incompressible flow   fluid flow in which density changes can be neglected. *Compare with* **compressible flow**

inertia   resistance of a body to a change in its state of motion

infrared (IR) radiation   that portion of the electromagnetic (EM) spectrum lying between the optical (visible) and radio wavelengths

International System of units   *See* **SI unit system**

inviscid fluid   perfect fluid that has zero coefficient of viscosity

ion   atom or molecule that has lost or gained one or more electrons, so that the total number of electrons does not equal the number of protons

ionic bond   formed when one atom gives up at least one outer electron to another atom, creating a chemical bond–producing electrical attraction between the atoms

isotope   atoms of the same chemical element but with different numbers of neutrons in their nucleus

joule (J)   basic unit of energy or work in the SI unit system; 1 J = 0.2388 calorie = 0.00095 Btu

kelvin (K)   SI unit of absolute thermodynamic temperature

kinetic energy (KE)   energy due to motion

lepton   fundamental particle of matter that does not participate in strong interactions; in the standard model, the three charged leptons are the electron (e), the muon (μ), and the tau (τ) particle; the three neutral leptons are the electron neutrino (v_e), the muon neutrino (v_μ), and the tau neutrino (v_τ). A corresponding set of antiparticles also exists. *See also* **standard model**

light-year (ly)   distance light travels in one year; 1 ly $\approx 5.88 \times 10^{12}$ miles (9.46 $\times 10^{12}$ km)

liquid   state of matter characterized as a relatively incompressible flowing fluid that maintains an essentially constant volume but assumes the shape of its container

liter (l or L)   SI unit of volume; 1 L = 0.264 gal

magnet   material or device that exhibits magnetic properties capable of causing the attraction or repulsion of another magnet or the attraction of certain ferromagnetic materials such as iron

manufacturing   process of transforming raw material(s) into a finished product, especially in large quantities

mass (m)   property that describes how much material makes up an object and gives rise to an object's inertia

mass number   *See* **relative atomic mass**

mass spectrometer   instrument that measures relative atomic masses and relative abundances of isotopes

material   tangible substance (chemical, biological, or mixed) that goes into the makeup of a physical object

mechanics   branch of physics that deals with the motions of objects

melting point   temperature at which a substance experiences a change from the solid state to the liquid state at a specified pressure; at this temperature, the solid and liquid states of a substance can coexist in equilibrium. *Synonymous with* **freezing point**

metallic bond   chemical bond created as many atoms of a metallic substance share the same electrons

meter (m)   fundamental SI unit of length; 1 meter = 3.281 feet. British spelling *metre*

metric system   *See* **SI unit system**

metrology   science of dimensional measurement; sometimes includes the science of weighing

microwave (radiation)   comparatively short-wavelength electromagnetic (EM) wave in the radio frequency portion of the EM spectrum

mirror matter   *See* **antimatter**

mixture   a combination of two or more substances, each of which retains its own chemical identity

molarity (M)   concentration of a solution expressed as moles of solute per kilogram of solvent

mole (mol)   SI unit of the amount of a substance; defined as the amount of substance that contains as many elementary units as there are atoms in 0.012 kilograms of carbon-12, a quantity known as Avogadro's number (N_A), which has a value of about 6.022×10^{23} molecules/mole

molecule   smallest amount of a substance that retains the chemical properties of the substance; held together by chemical bonds, a molecule can consist of identical atoms or different types of atoms

monomer   substance of relatively low molecular mass; any of the small molecules that are linked together by covalent bonds to form a polymer

natural material   material found in nature, such as wood, stone, gases, and clay

neutrino (ν)   lepton with no electric charge and extremely low (if not zero) mass; three known types of neutrinos are the electron neutrino (ν_e), the muon neutrino (ν_μ), and the tau neutrino (ν_τ). *See also* **lepton**

neutron (n)   an uncharged elementary particle found in the nucleus of all atoms except ordinary hydrogen. Within the standard model, the neutron is a baryon with zero electric charge consisting of two down (d) quarks and one up (u) quark. *See also* **standard model**

newton (N)   The SI unit of force; 1 N = 0.2248 lbf

nuclear chain reaction   occurs when a fissionable nuclide (such as plutonium-239) absorbs a neutron, splits (or fissions), and releases several neutrons along with energy. A fission chain reaction is self-sustaining when (on average) at least one released neutron per fission event survives to create another fission reaction

nuclear energy   energy released by a nuclear reaction (fission or fusion) or by radioactive decay

nuclear radiation   particle and electromagnetic radiation emitted from atomic nuclei as a result of various nuclear processes, such as radioactive decay and fission

nuclear reaction   reaction involving a change in an atomic nucleus, such as fission, fusion, neutron capture, or radioactive decay

nuclear reactor   device in which a fission chain reaction can be initiated, maintained, and controlled

nuclear weapon   precisely engineered device that releases nuclear energy in an explosive manner as a result of nuclear reactions involving fission, fusion, or both

nucleon   constituent of an atomic nucleus; a proton or a neutron

nucleus   (plural: nuclei) small, positively charged central region of an atom that contains essentially all of its mass. All nuclei contain both protons and neutrons except the nucleus of ordinary hydrogen, which consists of a single proton

nuclide   general term applicable to all atomic (isotopic) forms of all the elements; nuclides are distinguished by their atomic number, relative mass number (atomic mass), and energy state

ohm (Ω)   SI unit of electrical resistance

oxidation   chemical reaction in which oxygen combines with another substance, and the substance experiences one of three processes: (1) the gaining of oxygen, (2) the loss of hydrogen, or (3) the loss of electrons. In these reactions, the substance being "oxidized" loses electrons and forms positive ions. *Compare with* **reduction**

oxidation-reduction (redox) reaction   chemical reaction in which electrons are transferred between species or in which atoms change oxidation number

particle   minute constituent of matter, generally one with a measurable mass

pascal (Pa)   SI unit of pressure; 1 Pa = 1 N/m² = 0.000145 psi

Pascal's principle   when an enclosed (static) fluid experiences an increase in pressure, the increase is transmitted throughout the fluid; the physical principle behind all hydraulic systems

Pauli exclusion principle   postulate that no two electrons in an atom can occupy the same quantum state at the same time; also applies to protons and neutrons

perfect fluid   hypothesized fluid primarily characterized by a lack of viscosity and usually by incompressibility

perfect gas law   *See* **ideal gas law**

periodic table   list of all the known elements, arranged in rows (periods) in order of increasing atomic numbers and columns (groups) by similar physical and chemical characteristics

phase   one of several different homogeneous materials present in a portion of matter under study; the set of states of a large-scale (macroscopic) physical system having relatively uniform physical properties and chemical composition

phase transition   *See* **change of state**

photon   A unit (or particle) of electromagnetic radiation that carries a quantum (packet) of energy that is characteristic of the particular radiation. Photons travel at the speed of light and have an effective momentum, but no mass or electrical charge. In the standard model, a photon is the carrier particle of electromagnetic radiation

photovoltaic cell   *See* **solar cell**

physical property   characteristic quality of a substance that can be measured or demonstrated without changing the composition or chemical identity of the substance, such as temperature and density. *Compare with* **chemical property**

Planck's constant (h)   fundamental physical constant describing the extent to which quantum mechanical behavior influences nature. Equals the ratio of a photon's energy (E) to its frequency (v), namely: $h = E/v = 6.626 \times 10^{-34}$ J-s (6.282×10^{-37} Btu-s). *See also* **uncertainty principle**

plasma   electrically neutral gaseous mixture of positive and negative ions; called the fourth state of matter

plastic deformation   permanent change in size or shape of a solid due to an applied force (stress)

plasticity   tendency of a loaded body to assume a (deformed) state other than its original state when the load is removed

plastics   synthesized family of organic (mainly hydrocarbon) polymer materials used in nearly every aspect of modern life

pneumatic   operated, moved, or effected by a pressurized gas (typically air) that is used to transmit energy

polymer   very large molecule consisting of a number of smaller molecules linked together repeatedly by covalent bonds, thereby forming long chains

positron (e$^+$ or β$^+$)   elementary antimatter particle with the mass of an electron but charged positively

pound-force (lbf)   basic unit of force in the American customary system; 1 lbf = 4.448 N

pound-mass (lbm)   basic unit of mass in the American customary system; 1 lbm = 0.4536 kg

power   rate with respect to time at which work is done or energy is transformed or transferred to another location; 1 hp = 550 ft-lb$_{force}$/s = 746 W

pressure (P)   the normal component of force per unit area exerted by a fluid on a boundary; 1 psi = 6,895 Pa

product   substance produced by or resulting from a chemical reaction

proton (p)   stable elementary particle with a single positive charge. In the the standard model, the proton is a baryon with an electric charge of +1; it consists of two up (u) quarks and one down (d) quark. *See also* **standard model**

quantum mechanics   branch of physics that deals with matter and energy on a very small scale; physical quantities are restricted to discrete values and energy to discrete packets called quanta

quark   fundamental matter particle that experiences strong-force interactions. The six flavors of quarks in order of increasing mass are up (u), down (d), strange (s), charm (c), bottom (b), and top (t)

radiation heat transfer   The transfer of heat by electromagnetic radiation that arises due to the temperature of a body; can takes place in and through a vacuum

radioactive isotope   unstable isotope of an element that decays or disintegrates spontaneously, emitting nuclear radiation; also called radioisotope

radioactivity   spontaneous decay of an unstable atomic nucleus, usually accompanied by the emission of nuclear radiation, such as alpha particles, beta particles, gamma rays, or neutrons

radio frequency (RF)   a frequency at which electromagnetic radiation is useful for communication purposes; specifically, a frequency above 10,000 hertz (Hz) and below 3×10^{11} Hz

rankine (R)   American customary unit of absolute temperature. *See also* **kelvin (K)**

reactant   original substance or initial material in a chemical reaction

reduction   portion of an oxidation-reduction (redox) reaction in which there is a gain of electrons, a gain in hydrogen, or a loss of oxygen. *See also* **oxidation-reduction (redox) reaction**

relative atomic mass (A)   total number of protons and neutrons (nucleons) in the nucleus of an atom. Previously called *atomic mass* or *atomic mass number. See also* **atomic mass unit**

residual electromagnetic effect   force between electrically neutral atoms that leads to the formation of molecules

residual strong interaction   interaction responsible for the nuclear binding force—that is, the strong force holding hadrons (protons and neutrons) together in the atomic nucleus. *See also* **strong force**

resilience   property of a material that enables it to return to its original shape and size after deformation

resistance (R)   the ratio of the voltage (V) across a conductor to the electric current (I) flowing through it

scientific notation   A method of expressing powers of 10 that greatly simplifies writing large numbers; for example, $3 \times 10^6 = 3,000,000$

SI unit system   international system of units (the metric system), based upon the meter (m), kilogram (kg), and second (s) as the fundamental units of length, mass, and time, respectively

solar cell (photovoltaic cell)   a semiconductor direct energy conversion device that transforms sunlight into electric energy

solid   state of matter characterized by a three-dimensional regularity of structure; a solid is relatively incompressible, maintains a fixed volume, and has a definitive shape

solution When scientists dissolve a substance in a pure liquid, they refer to the dissolved substance as the *solute* and the host pure liquid as the *solvent*. They call the resulting intimate mixture the solution

spectroscopy study of spectral lines from various atoms and molecules; emission spectroscopy infers the material composition of the objects that emitted the light; absorption spectroscopy infers the composition of the intervening medium

speed of light *(c)* speed at which electromagnetic radiation moves through a vacuum; regarded as a universal constant equal to 186,283.397 mi/s (299,792.458 km/s)

stable isotope isotope that does not undergo radioactive decay

standard model contemporary theory of matter, consisting of 12 fundamental particles (six quarks and six leptons), their respective antiparticles, and four force carriers (gluons, photons, W bosons, and Z bosons)

state of matter form of matter having physical properties that are quantitatively and qualitatively different from other states of matter; the three more common states on Earth are solid, liquid, and gas

steady state condition of a physical system in which parameters of importance (fluid velocity, temperature, pressure, etc.) do not vary significantly with time

strain the change in the shape or dimensions (volume) of an object due to applied forces; longitudinal, volume, and shear are the three basic types of strain

stress applied force per unit area that causes an object to deform (experience strain); the three basic types of stress are compressive (or tensile) stress, hydrostatic pressure, and shear stress

string theory theory of quantum gravity that incorporates Einstein's general relativity with quantum mechanics in an effort to explain space-time phenomena on the smallest imaginable scales; vibrations of incredibly tiny stringlike structures form quarks and leptons

strong force In the standard model, the fundamental force between quarks and gluons that makes them combine to form hadrons, such as protons and neutrons; also holds hadrons together in a nucleus. *See also* **standard model**

subatomic particle any particle that is small compared to the size of an atom

sublimation   direct transition of a material from the solid state to the gaseous (vapor) state without passing through the liquid phase. *Compare* **deposition**

superconductivity   the ability of a material to conduct electricity without resistance at a temperature above absolute zero

temperature (T)   thermodynamic property that serves as a macroscopic measure of atomic and molecular motions within a substance; heat naturally flows from regions of higher temperature to regions of lower temperature

tension   condition when applied external forces pull atoms of a material farther apart. *Compare* **compression**

thermal conductivity (k)   intrinsic physical property of a substance; a material's ability to conduct heat as a consequence of molecular motion

thermodynamics   branch of science that treats the relationships between heat and energy, especially mechanical energy

thermodynamic system   collection of matter and space with boundaries defined in such a way that energy transfer (as work and heat) from and to the system across these boundaries can be easily identified and analyzed

thermometer   instrument or device for measuring temperature

toughness   ability of a material (especially a metal) to absorb energy and deform plastically before fracturing

transmutation   transformation of one chemical element into a different chemical element by a nuclear reaction or series of reactions

transuranic element (isotope)   human-made element (isotope) beyond uranium on the periodic table

ultraviolet (UV) radiation   portion of the electromagnetic spectrum that lies between visible light and X-rays

uncertainty principle   Heisenberg's postulate that places quantum-level limits on how accurately a particle's momentum *(p)* and position *(x)* can be simultaneously measured. Planck's constant (h) expresses this uncertainty as $\Delta x \times \Delta p \geq h/2\pi$

U.S. customary system of units   *See* **American customary system of units**

vacuum   relative term used to indicate the absence of gas or a region in which there is a very low gas pressure

valence electron   electron in the outermost shell of an atom

van der Waals force   generally weak interatomic or intermolecular force caused by polarization of electrically neutral atoms or molecules

vapor   gaseous state of a substance

velocity   vector quantity describing the rate of change of position; expressed as length per unit of time

velocity of light (*c*)   *See* **speed of light**

viscosity   measure of the internal friction or flow resistance of a fluid when it is subjected to shear stress

volatile   solid or liquid material that easily vaporizes; volatile material has a relatively high vapor pressure at normal temperatures

volt (V)   SI unit of electric potential difference

volume (V)   space occupied by a solid object or a mass of fluid (liquid or confined gas)

watt (W)   SI unit of power (work per unit time); $1 \text{ W} = 1 \text{ J/s} = 0.00134 \text{ hp} = 0.737 \text{ ft-lb}_{force}/s$

wavelength (λ)   the mean distance between two adjacent maxima (or minima) of a wave

weak force   fundamental force of nature responsible for various types of radioactive decay

weight (*w*)   the force of gravity on a body; on Earth, product of the mass (m) of a body times the acceleration of gravity (*g*), namely $w = m \times g$

work (W)   energy expended by a force acting though a distance. *Compare* **heat**

X-ray   penetrating form of electromagnetic (EM) radiation that occurs on the EM spectrum between ultraviolet radiation and gamma rays

Further Resources

BOOKS

Allcock, Harry R. *Introduction to Materials Chemistry.* New York: John Wiley & Sons, 2008. A college-level textbook that provides a basic treatment of the principles of chemistry upon which materials science depends.

Angelo, Joseph A., Jr. *Nuclear Technology.* Westport, Conn.: Greenwood Press, 2004. The book provides a detailed discussion of both military and civilian nuclear technology and includes impacts, issues, and future advances.

———. *Encyclopedia of Space and Astronomy.* New York: Facts On File, 2006. Provides a comprehensive treatment of major concepts in astronomy, astrophysics, planetary science, cosmology, and space technology.

Ball, Philip. *Designing the Molecular World: Chemistry at the Frontier.* Princeton, N.J.: Princeton University Press, 1996. Discusses many recent advances in modern chemistry, including nanotechnology and superconductor materials.

———. *Made to Measure: New Materials for the 21st Century.* Princeton, N.J.: Princeton University Press, 1998. Discusses how advanced new materials can significantly influence life in the 21st century.

Bensaude-Vincent, Bernadette, and Isabelle Stengers. *A History of Chemistry.* Cambridge, Mass.: Harvard University Press, 1996. Describes how chemistry emerged as a science and its impact on civilization.

Callister, William D., Jr. *Materials Science and Engineering: An Introduction.* 8th ed. New York: John Wiley & Sons, 2010. Intended primarily for engineers, technically knowledgeable readers will also benefit from this book's introductory treatment of metals, ceramics, polymers, and composite materials.

Charap, John M. *Explaining the Universe: The New Age of Physics.* Princeton, N.J.: Princeton University Press, 2004. Discusses the important discoveries in physics during the 20th century that are influencing civilization.

Close, Frank, et al. *The Particle Odyssey: A Journey to the Heart of the Matter.* New York: Oxford University Press, 2002. A well-illustrated and enjoyable tour of the subatomic world.

Cobb, Cathy, and Harold Goldwhite. *Creations of Fire: Chemistry's Lively History from Alchemy to the Atomic Age.* New York: Plenum Press, 1995. Uses historic circumstances and interesting individuals to describe the emergence of chemistry as a scientific discipline.

Feynman, Richard P. *QED: The Strange Theory of Light and Matter.* Princeton, N.J.: Princeton University Press, 2006. Written by an American Nobel laureate, addresses several key topics in modern physics.

Gordon, J. E. *The New Science of Strong Materials or Why You Don't Fall Through the Floor.* Princeton, N.J.: Princeton University Press, 2006. Discusses the science of structural materials in a manner suitable for both technical and lay audiences.

Hill, John W., and Doris K. Kolb. *Chemistry for Changing Times.* 11th ed. Upper Saddle River, N.J.: Pearson Prentice Hall, 2007. Readable college-level textbook that introduces all the basic areas of modern chemistry.

Krebs, Robert E. *The History and Use of Our Earth's Chemical Elements: A Reference Guide.* 2nd ed. Westport, Conn.: Greenwood Press, 2006. Provides a concise treatment of each of the chemical elements.

Levere, Trevor H. *Transforming Matter: A History of Chemistry from Alchemy to the Buckyball.* Baltimore: Johns Hopkins University Press, 2001. Provides an understandable overview of the chemical sciences from the early alchemists through modern times.

Lutgens, Frederick K., and Edward J. Tarbuck. *The Atmosphere: An Introduction to Meteorology.* 10th ed. Upper Saddle River, N.J.: Pearson Prentice Hall, 2007. Readable college-level textbook that discusses the atmosphere, meteorology, climate, and the physical properties of air.

Mackintosh, Ray, et al. *Nucleus: A Trip into the Heart of Matter.* Baltimore: Johns Hopkins University Press, 2001. Provides a technical though readable explanation of how modern scientists developed their current understanding of the atomic nucleus and the standard model.

Nicolaou, K. C., and Tamsyn Montagnon. *Molecules that Changed the World.* New York: John Wiley & Sons, 2008. Provides an interesting treatment of such important molecules as aspirin, camphor, glucose, quinine, and morphine.

Scerri, Eric R. *The Periodic Table: Its Story and Its Significance.* New York: Oxford University Press, 2007. Provides a detailed look at the periodic table and its iconic role in the practice of modern science.

Smith, William F., and Javad Hashemi. *Foundations of Materials Science and Engineering.* 5th ed. New York: McGraw-Hill, 2006. Provides scientists and engineers of all disciplines an introduction to materials science, including metals, ceramics, polymers, and composite materials. Technically knowledgeable laypersons will find the treatment of specific topics such as biological materials useful.

Strathern, Paul. *Mendeleyev's Dream: The Quest for the Elements.* New York: St. Martin's Press, 2001. Describes the intriguing history of chemistry from the early Greek philosophers to the 19th-century Russian chemist Dmitri Mendeleyev.

Thrower, Peter, and Thomas Mason. *Materials in Today's World.* 3rd ed. New York: McGraw-Hill Companies, 2007. Provides a readable introductory treatment of modern materials science, including biomaterials and nanomaterials.

Trefil, James, and Robert M. Hazen. *Physics Matters: An Introduction to Conceptual Physics.* New York: John Wiley & Sons, 2004. Highly-readable introductory college-level textbook that provides a good overview of physics from classical mechanics to relativity and cosmology. Laypersons will find the treatment of specific topics useful and comprehendible.

Zee, Anthony. *Quantum Field Theory in a Nutshell.* Princeton, N.J.: Princeton University Press, 2003. A reader-friendly treatment of the generally complex and profound physical concepts that constitute quantum field theory.

WEB SITES

To help enrich the content of this book and to make your investigation of matter more enjoyable, the following is a selective list of recommended Web sites. Many of the sites below will also lead to other interesting science-related locations on the Internet. Some sites provide unusual science learning opportunities (such as laboratory simulations) or in-depth educational resources.

American Chemical Society (ACS) is a congressionally chartered independent membership organization that represents professionals at all degree levels and in all fields of science involving chemistry. The ACS Web site includes educational resources for high school and college students. Available online. URL: http://portal.acs.org/portal/acs/corg/content. Accessed on February 12, 2010.

American Institute of Physics (AIP) is a not-for-profit corporation that promotes the advancement and diffusion of the knowledge of physics and its applications to human welfare. This Web site offers an enormous quantity of fascinating information about the history of physics from ancient Greece up to the present day. Available online. URL: http://www.aip.org/aip/. Accessed on February 12, 2010.

Chandra X-ray Observatory (CXO) is a space-based NASA astronomical observatory that observes the universe in the X-ray portion of the electromagnetic spectrum. This Web site contains contemporary information and educational materials about astronomy, astrophysics, and cosmology, including topics such as black holes, neutron stars, dark matter, and dark energy. Available online. URL: http://www.chandra.harvard.edu/. Accessed on February 12, 2010.

The ChemCollective is an online resource for learning about chemistry. Through simulations developed by the Department of Chemistry of Carnegie Mellon University (with funding from the National Science Foundation), a person gets the chance to safely mix chemicals without worrying about accidentally spilling them. Available online. URL: http://www.chemcollective.org/vlab/vlab.php. Accessed on February 12, 2010.

Chemical Heritage Foundation (CHF) maintains a rich and informative collection of materials that describe the history and heritage of the chemical and molecular sciences, technologies, and industries. Available online. URL: http://www.chemheritage.org/. Accessed on February 12, 2010.

Department of Defense (DOD) is responsible for maintaining armed forces of sufficient strength and technology to protect the United States and its citizens from all credible foreign threats. This Web site serves as an efficient access point to activities within the DOD, including those taking place within each of the individual armed services: the U.S. Army, U.S. Navy, U.S. Air Force, and U.S. Marines. As part of national security, the DOD sponsors a large amount of research and development, including activities in materials science, chemistry, physics, and nanotechnology. Available online. URL: http://www.defenselink.mil/. Accessed on February 12, 2010.

Department of Energy (DOE) is the single largest supporter of basic research in the physical sciences in the federal government of the United States. Topics found on this Web site include materials sciences, nanotechnology, energy sciences, chemical science, high-energy physics, and nuclear physics. The Web site also includes convenient links to all of the DOE's national laboratories. Available online. URL: http://energy.gov/. Accessed on February 12, 2010.

Fermi National Accelerator Laboratory (Fermilab) performs research that advances the understanding of the fundamental nature of matter and energy. Fermilab's Web site contains contemporary information about particle physics, the standard model, and the impact of particle physics on society. Available online. URL: http://www.fnal.gov/. Accessed on February 12, 2010.

Hubble Space Telescope (HST) is a space-based NASA observatory that has examined the universe in the (mainly) visible portion of the electromagnetic spectrum. This Web site contains contemporary information and educational materials about astronomy, astrophysics, and cosmology, including topics such as black holes, neutron stars, dark matter, and dark energy. Available online. URL: http://hubblesite.org/. Accessed on February 12, 2010.

Institute and Museum of the History of Science in Florence, Italy, offers a special collection of scientific instruments (some viewable online), including those used by Galileo Galilei. Available online. URL: http://www.imss.fi.it/. Accessed on February 12, 2010.

International Union of Pure and Applied Chemistry (IUPAC) is an international nongovernmental organization that fosters worldwide communications in the chemical sciences and in providing a common language for chemistry that unifies the industrial, academic, and public sectors. Available online. URL: http://www.iupac.org/. Accessed on February 12, 2010.

National Aeronautics and Space Administration (NASA) is the civilian space agency of the U.S. government and was created in 1958 by an act of Congress. NASA's overall mission is to direct, plan, and conduct American civilian (including scientific) aeronautical and space activities for peaceful purposes. Available online. URL: http://www.nasa.gov/. Accessed on February 12, 2010.

National Institute of Standards and Technology (NIST) is an agency of the U.S. Department of Commerce that was founded in 1901 as the nation's first federal physical science research laboratory. The NIST Web site includes contemporary information about many areas of science and engineering, including analytical chemistry, atomic and molecular physics, biometrics, chemical and crystal structure, chemical kinetics, chemistry, construction, environmental data, fire, fluids, material properties, physics, and thermodynamics. Available online. URL: http://www.nist.gov/index.html. Accessed on February 12, 2010.

National Oceanic and Atmospheric Administration (NOAA) was established in 1970 as an agency within the U.S. Department of Commerce to ensure the safety of the general public from atmospheric phenomena and to provide the public with an understanding of Earth's environment and resources. Available online. URL: http://www.noaa.gov/. Accessed on February 12, 2010.

NEWTON: Ask a Scientist is an electronic community for science, math, and computer science educators and students sponsored by the Argonne National Laboratory (ANL) and the U.S. Department of Energy's Office of Science Education. This Web site provides access to a fascinating list of questions and answers involving the following disciplines/topics: astronomy, biology, botany, chemistry, computer science, Earth science, engineering, environmental science, general science, materials science, mathematics, molecular biology, physics, veterinary, weather, and zoology. Available online. URL: http://www.newton.dep.anl.gov/archive.htm. Accessed on February 12, 2010.

Nobel Prizes in Chemistry and Physics. This Web site contains an enormous amount of information about all the Nobel Prizes awarded in physics and chemistry, as well as complementary technical information. Available online. URL: http://nobelprize.org/. Accessed on February 12, 2010.

Periodic Table of Elements. An informative online periodic table of the elements maintained by the Chemistry Division of the Department of Energy's Los Alamos National Laboratory (LANL). Available online. URL: http://periodic.lanl.gov/. Accessed on February 12, 2010.

PhET Interactive Simulations is an ongoing effort by the University of Colorado at Boulder (under National Science Foundation sponsorship) to provide a comprehensive collection of simulations to enhance science learning. The major science categories include physics, chemistry, Earth science, and biology. Available online. URL: http://phet.colorado.edu/index.php. Accessed on February 12, 2010.

ScienceNews is the online version of the magazine of the Society for Science and the Public. Provides insights into the latest scientific achievements and discoveries. Especially useful are the categories Atom and Cosmos, Environment, Matter and Energy, Molecules, and Science and Society.

Available online. URL: http://www.sciencenews.org/. Accessed on February 12, 2010.

The Society on Social Implications of Technology (SSIT) of the Institute of Electrical and Electronics Engineers (IEEE) deals with such issues as the environmental, health, and safety implications of technology; engineering ethics; and the social issues related to telecommunications, information technology, and energy. Available online. URL: http://www.ieeessit.org/. Accessed on February 12, 2010.

Spitzer Space Telescope (SST) is a space-based NASA astronomical observatory that observes the universe in the infrared portion of the electromagnetic spectrum. This Web site contains contemporary information and educational materials about astronomy, astrophysics, and cosmology, including the infrared universe, star and planet formation, and infrared radiation. Available online. URL: http://www.spitzer.caltech.edu/. Accessed on February 12, 2010.

Thomas Jefferson National Accelerator Facility (Jefferson Lab) is a U.S. Department of Energy–sponsored laboratory that conducts basic research on the atomic nucleus at the quark level. The Web site includes basic information about the periodic table, particle physics, and quarks. Available online. URL: http://www.jlab.org/. Accessed on February 12, 2010.

United States Geological Survey (USGS) is the agency within the U.S. Department of the Interior that serves the nation by providing reliable scientific information needed to describe and understand Earth, minimize the loss of life and property from natural disasters, and manage water, biological, energy, and mineral resources. The USGS Web site is rich in science information, including the atmosphere and climate, Earth characteristics, ecology and environment, natural hazards, natural resources, oceans and coastlines, environmental issues, geologic processes, hydrologic processes, and water resources. Available online. URL: http://www.usgs.gov/. Accessed on February 12, 2010.

Index

Italic page numbers indicate illustrations.